1 MONTH OF
FREE
READING

at
www.ForgottenBooks.com

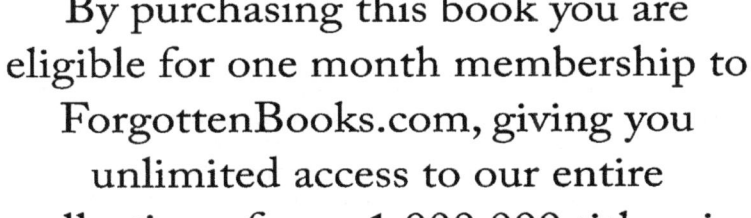

By purchasing this book you are eligible for one month membership to ForgottenBooks.com, giving you unlimited access to our entire collection of over 1,000,000 titles via our web site and mobile apps.

To claim your free month visit:
www.forgottenbooks.com/free907914

ISBN 978-0-265-90725-2
PIBN 10907914

Forgotten Books is a registered trademark of FB &c Ltd.
Copyright © 2018 FB &c Ltd.
FB &c Ltd, Dalton House, 60 Windsor Avenue, London, SW19 2RR.
Company number 08720141. Registered in England and Wales.

te for Historical Microreproductions / Institut canadien de microreproductions historiques

1997

Bibliographic Notes / Notes techniques et bibliographiques

o obtain the best original
atures of this copy which
e, which may alter any of
luction, or which may
al method of filming are

L'Institut a microfilmé le meilleur exemplaire qu'il lui a été possible de se procurer. Les détails de cet exemplaire qui sont peut-être uniques du point de vue bibliographique, qui peuvent modifier une image reproduite, ou qui peuvent exiger une modification dans la méthode normale de filmage sont indiqués ci-dessous.

- [ ] Coloured pages / Pages de couleur
- [ ] Pages damaged / Pages endommagées
- [ ] Pages restored and/or laminated / Pages restaurées et/ou pelliculées
- [x] Pages discoloured, stained or foxed / Pages décolorées, tachetées ou piquées
- [ ] Pages detached / Pages détachées
- [x] Showthrough / Transparence
- [x] Quality of print varies / Qualité inégale de l'impression
- [ ] Includes supplementary material / Comprend du matériel supplémentaire
- [ ] Pages wholly or partially obscured by errata slips, tissues, etc., have been refilmed to ensure the best possible image / Les pages totalement ou partiellement obscurcies par un feuillet d'errata, une pelure, etc., ont été filmées à nouveau de façon à obtenir la meilleure image possible.
- [ ] Opposing pages with varying colouration or discolourations are filmed twice to ensure the best possible image / Les pages s'opposant ayant des colorations variables ou des décolorations sont filmées deux fois afin d'obtenir la meilleure image possible.

e

aminated /
ou pelliculée

tre de couverture manque

géographiques en couleur

han blue or black) /
tre que bleue ou noire)

lustrations /
ns en couleur

al /
ments

shadows or distortion along
ure serrée peut causer de
sion le long de la marge

ng restorations may appear
r possible, these have been
e peut que certaines pages
ors d'une restauration
te, mais, lorsque cela était
t as été filmées

uced thanks

a

st quality
d legibility
ith the

e are filmed
ding on
ted impres-
iate. All
ning on the
impres-
a printed

ofiche
g "CON-
"END"),

ed at
rge to be
filmed
r, left to
es as
trate the

L'exemplaire filmé fut reproduit grâce à la
générosité de:

Bibliothèque nationale du Canada

Les images suivantes ont été reproduites avec le
plus grand soin, compte tenu de la condition et
de la netteté de l'exemplaire filmé, et en
conformité avec les conditions du contrat de
filmage.

Les exemplaires originaux dont la couverture en
papier est imprimée sont filmés en commençant
par le premier plat et en terminant soit par la
dernière page qui comporte une empreinte
d'impression ou d'illustration, soit par le second
plat, selon le cas. Tous les autres exemplaires
originaux sont filmés en commençant par la
première page qui comporte une empreinte
d'impression ou d'illustration et en terminant par
la dernière page qui comporte une telle
empreinte.

Un des symboles suivants apparaîtra sur la
dernière image de chaque microfiche, selon le
cas: le symbole ➔ signifie "A SUIVRE", le
symbole ▽ signifie "FIN".

Les cartes, planches, tableaux, etc., peuvent être
filmés à des taux de réduction différents.
Lorsque le document est trop grand pour être
reproduit en un seul cliché, il est filmé à partir
de l'angle supérieur gauche, de gauche à droite,
et de haut en bas, en prenant le nombre
d'images nécessaire. Les diagrammes suivants
illustrent la méthode.

3   1

MICROCOPY RESOLUTION TEST CHART

(ANSI and ISO TEST CHART No. 2)

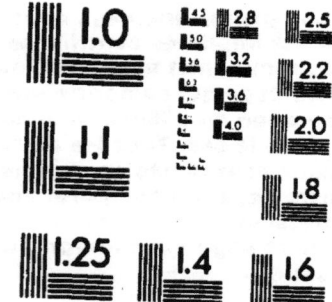

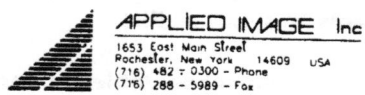

APPLIED IMAGE Inc
1653 East Main Street
Rochester, New York   14609   USA
(716) 482 - 0300 - Phone
(716) 288 - 5989 - Fax

CANADA

DEPARTMENT OF MINES

GEOLOGICAL SURVEY BRANCH.

Hon. W. Templeman, Minister; A. P. Low, Deputy Minister:
R. W. Brock, Director.

MEMOIR Nᵒ· 5

PRELIMINARY MEMOIR

ON THE

LEWES AND NORDENSKIÖLD RIVERS COAL DISTRICT

YUKON TERRITORY

BY

D. D. CAIRNES

OTTAWA
GOVERNMENT PRINTING BUREAU
1910

No. 1101

Plate I.

Breaking up of the ice on Lake Laberge, June 5, 1897.

CANADA

DEPARTMENT OF MINES

GEOLOGICAL SURVEY BRANCH.

Hon. W. Templeman, Minister: A. P. Low, Deputy Minister.
R. W. Brock, Director.

MEMOIR No. 5

PRELIMINARY MEMOIR

ON THE

LEWES AND NORDENSKIÖLD RIVERS COAL DISTRICT

YUKON TERRITORY

BY

D. D. CAIRNES

OTTAWA
GOVERNMENT PRINTING BUREAU
1910

To R. W. Brock, Esq.,
        Director Geological Survey,
                Department of Mir.

Sir,—I beg to submit the following 'Preliminary Memoir on the
Lewes and Nordenskiöld Rivers Coal District,' accompanying which,
are two topographical and geological maps of the Tantalus Coal
Area and the Braeburn-Kynocks Coal Area, respectively.

                I have the honour to be, sir,
                        Your obedient servant,

                                (Signed) D. D. CAIRNES.

May 30, 1909.

CONTENTS.

5

APPENDIX II.

APPENDIX III.

ILLUSTRATIONS.

PHOTOGRAPHS.

MAPS.

PRELIMINARY MEMOIR

ON THE

LEWES AND NORDENSKIÖLD RIVERS COAL DISTRICT,

YUKON TERRITORY

BY

D. D. CAIRNES.

--- --

INTRODUCTORY.

The recent development of the Whitehorse copper ore deposits having reached such magnitude and importance that the installation of a local smelter is under serious consideration, it has been deemed advisable to obtain reliable information as to the nearest accessible point to Whitehorse where metallurgical coking coal can be procured, and at the same time to determine the extent of the coal-bearing formations of the district; for, hitherto, coal in the Whitehorse copper belt region has only been required for fuel on the river steamers plying between Whitehorse and Dawson, and for domestic purposes in these towns.

Fossil fuel of good quality has been known to exist on the Lewes river—at Tantalus, Tantalus butte, and the Five Fingers mine— for over twenty-two years, also at places along the Nordenskiöld river, which enters the Lewes river at Tantalus. The coal from Tantalus and Five Fingers gave very satisfactory caloric and coking results when analysed and tested in the chemical laboratory of the Mines Branch of the Department of Mines. The occurrence of coal has also been reported as existing at numerous places between Whitehorse and Tantalus: along Lake Laberge, and Lewes river; also along the tributaries of the latter, namely, Teslin and Big Salmon rivers. Hence, the writer was instructed to explore the region north of Whitehorse, with a view to mapping and geologically examining those portions found to be coal-bearing, and to

7

determine, particularly, the nearest points accessible to Whitehorse at which coal of the same geological horizon as that at Tantalus could be obtained.

Field work was commenced in the vicinity of Tantalus, during the season of 1907, and a map projected to include the Tantalus and Five Fingers mines and Tantalus butte. During the summer of 1908, the area mapped was extended to the south, to include the most interesting portions of the district. During the former season map work was commenced also at the south end of Lake Laberge and extended west, cross-cutting the various formations, and in 1908 was continued so as to correct the most southerly extension of the Tantalus coal measures, north of Whitehorse. There is a strip still unmapped between these surveyed areas, but, in the meantime, it is considered that the most important parts of the district have been covered; a rapid geological reconnaissance having been made from Whitehorse to Tantalus to ascertain the areas containing the coal-bearing formations.

A triangulation was carried through the areas mapped, four bases being measured. The topography was done by means of plane-table and photo-topographic method. The former, only, was employed during the season of 1907, but both were used the summer following.

The Canadian-Alaskan Boundary surveyors had run a line of levels from sea-line at Skagway over the White Pass and Yukon railway, to Whitehorse, and thence along the Whitehorse-Dawson wagon-road to the north of Tantalus—these levels to be eventually carried to some point on the 141st parallel east of Dawson. A level for the levels was obtained from this survey, which passed through both the areas being mapped. The latitude and longitude of a point at Tantalus had already been determined by the Astronomical Department, and this position was used as a starting point for our work. Contour topographical and geological maps to accompany this report have been prepared on a scale of 2 miles to 1 inch, with contour intervals of 250 feet.

During both seasons Mr. H. Matheson acted as topographer, and performed the greater part of this portion of the work. During the season of 1908 I was assisted in geology by Dr. O. Stutzer, lecturer on geology at the Royal School of Mines, Freiberg, Saxony.

Location and Area.

The more southerly of the districts surveyed, the "Braeburn-Kynocks Coal Area," is slightly to the west of due north of Whitehorse, and 45 or 50 miles distant, in a direct line. It includes the southern end of Lake Laberge on the east, and extends thence 35 miles in a westerly direction, crossing the East and Middle branches of the Nordenskiöld river, known as Klusha creek, alongside which runs the Whitehorse-Dawson road, and Schwatka river. The map has an average width of 13 miles.

The more northerly map, that of the Tantalus Coal area, extends from north of the Five Fingers mine (which is 7 or 8 miles almost due north of Tantalus), about 43 miles in a southeasterly direction, with a width of 10 to 12 miles. To the south of Tantalus, this map includes the first range of hills along the west side of the Nordenskiöld river and extends to the east far enough to include the coal formations. Tantalus is 131 miles from Whitehorse, measured along the wagon road, and 105 miles in a direct line—in a direction a few degrees west of north. Between the areas included in the Tantalus and Braeburn-Kynocks maps, is an unmapped portion of country 14 or 15 miles wide.

Means of Communication.

Whitehorse is easily accessible. Steamers ply regularly between Skagway and both Vancouver and Seattle, which are 1,000 and 867 miles distant, respectively. From Skagway, Whitehorse—which is 111 miles distant—is reached by the White Pass and Yukon railway, which runs in a northerly direction, crossing the summit of the White Pass, on the Alaska-British Columbia boundary, 20.4 miles distant, and the 60th parallel on the British Columbia and Yukon boundary 51.4 miles distant from Skagway.

During the summer months, steamers leave Whitehorse for Dawson—a distance of 400 miles by river—several times a week. In the winter months, stages are driven regularly over the Whitehorse-Dawson road. So that points along navigable water below Whitehorse may also be considered quite accessible; but to reach places inland from these waterways, is, at present, both difficult and expensive.

History.

General History, and Previous Work.

The Lewes river was discovered by Robert Campbell in 1342, and named after the Chief Factor of the Hudson's Bay Co., John Lee Lewes. Previous to this time little was known of the southern Yukon Territory, and for a number of years afterwards it was visited only by a few traders or explorers. Previous to 1883, however, the Lewes and some of its tributaries had become well known to a number of miners and prospectors, and when, during that year, Lieut. Schwatka crossed the Chilcoot pass, and descended the Lewes river, he merely followed in their footsteps. But to him is due the credit of having made the first survey of the river: a survey which has since proved to be reasonably accurate as far as the main features are concerned. During the season of 1887, W. Ogilvie, D.L.S., carried out an instrumentally measured traverse of the route from the head of Lynn canal to the Lewes, and along the river to the 141st meridian. This may be considered the first accurate survey of the river, and is the best that has yet been made.

An interesting account of the early history of the Yukon has been written by Dr. G. M. Dawson,[1] including a description of the gold washings on the Lewes river and its tributaries the Big Salmon and Teslin rivers. These rivers have each yielded some gold, and it is possible that places may yet be found—particularly up the Teslin—which will pay for working.

During the years 1896 to 1908, when the rush to Dawson was at its height, great numbers of people, on their way to the Klondike, passed down the Lewes river, as well as over the Dalton trail, which follows the Nordenskiöld for several miles. Still—except for the comparatively small amount of gold obtained from the Lewes river, —the district considered in this report never attracted much attention until quite recently, when a few persons became interested in the coal deposits.

Dr. Dawson, in 1887, noted the occurrence of coal where the Five Fingers and Tantalus mines are now situated; but the first actual mining was commenced by Mr. C. J. Miller, at what is now known as Five Fingers mine. A few boat loads of coal were shipped

[1] Dr. G. M. Dawson. Report on Exploration in the Yukon district, N.W.T., and adjacent Northern Portions of British Columbia, 1887. Ann. Rep. Geol. Surv., Vol. III., Part B

to Dawson, but later the property was transferred to Mr. Geo. J. Milton, who organized the Five Fingers Coal Co., in 1905. Mr. Miller also caused work to be commenced at Tantalus, and shipped 370 tons of coal therefrom in 1904. Later, he located and prospected the Tantalus measures across the river on Tantalus butte, and has shown them to be, probably, the best so far discovered in the district.

Bibliography.

Dr. G. M. Dawson.—Report on an Exploration in the Yukon District, N.W.T., and Adjacent Northern Portion of British Columbia, 1887. Annual Report Geological Survey of Canada, Vol. III, Report B.

D. D. Cairnes.—Report on a Portion of the Conrad and Whitehorse Mining Districts. Geological Survey, Department of Mines, 1908.

D. D. Cairnes.—Exploration in a Portion of the Yukon, South of Whitehorse. Summary Report, Geological Survey of Canada, 1906.

D. D. Cairnes.—Report on a Portion of the Yukon Territory, chiefly between Whitehorse and Tantalus. Summary Report, Geological Survey, Department of Mines, 1907.

D. D. Cairnes.—Report on a Portion of the Yukon Territory west of the Lewes river, and between the Latitudes of Whitehorse and Tantalus. Summary Report, Geological Survey, Department of Mines, 1909.

Summary, and Conclusions.

There are two coal horizons of economic interest in this portion of the Yukon Territory: an upper horizon occurring near the top of the Tantalus conglomerate, to which belong the seams at the Tantalus mine and at Tantalus butte; and a lower horizon some distance below the Tantalus conglomerate which includes the seams of the Five Fingers mine; those to the west of the 69th mile-post from Whitehorse on the Whitehorse-Dawson road, and elsewhere. The seams of Tantalus butte and the Tantalus mine doubtless extend a number of miles to the north and south of these places respectively; for occasional outcrops of the coal were seen after the coal-bearing beds had been removed. but prospecting is rendered particularly difficult by the thick mantle of superficial material which covers the greater part of the district. With the exception of the first 2 or 3 miles, the coal-bearing sedimentary rocks to the south of Tantalus are, for the greater part, covered with recent basalts, basalt tuffs, etc.; so that, although in the 19 or 20 miles immediately south of Tantalus, there is believed to be a great amount of coal, it will, in most places, require very careful prospecting to find a suitable location for economic mining operations. South of this again, the coal formations have been removed by erosion for over 30 miles, to where a belt of the Tantalus conglomerate was discovered, traversing the district a few degrees in a northeast direction. Four miles east of the wagon-road, these beds became buried again under a great thickness of volcanics. To the northwest, however, they extend a number of miles, and possibly may be developed continuously to where they are found crossing the valley of the Hutshi river, about 30 miles above its confluence with the Norden-skiöld. The upper members of this terrane, including the upper coal horizon, have been eroded away, but the seams of the lower horizon were noted in several places.

Seeing that under present conditions in this district, coal can be considered economically accessible only when found alongside navigable water, attention was directed to the finding of workable deposits along the Lewes river, Lake Laberge, or their large tributaries. As yet, however, seams of importance have been discovered only along the Lewes, namely, at Tantalus butte, Tantalus, and Five Fingers mines. The Tantalus conglomerate was found along the west side of the river, a short distance above Tantalus, and also between Hootalinkwa and Big Salmon. It is probable that at the latter points coal may be found if the ground is carefully prospected, and would, if discovered, be of considerable value. The areas found along the Norden-skiöld constitute a valuable asset for the country, but cannot be utilized until a railway is built, or some means devised to use the coal in place. At present, Tantalus, and Tantalus butte are the closest known points present to Whitehorse at which workable coking coal is known to occur.

GENERAL CHARACTER OF DISTRICT.

Topography.

GENERAL ACCOUNT.

Regional.—The Lewes river, which joins the Pelly at Selkirk to form the Yukon—the fifth largest river on the North American continent—flows down the centre of the Yukon plateau, an intermontaine belt bounded on the west by the ranges of the Pacific Mountain system, and, on the east, by those of the Rocky Mountain system. The topography of this plateau province is, in a general way, characterized by distinct, often broad, and in places even flat-topped, elevated inter-stream-areas, separated by the valleys of a well developed drainage system, the valleys being usually broad and flat with gentle slopes, the inter-stream-areas marking the remnants of an upraised and subsequently dissected peneplain. Close to the Lewes and its larger tributaries, the erosion has been more active than farther away from those main waterways, and the plateau characteristics have been, to some extent, lost. Farther back, however, these features are more strongly emphasized. One traversing the valleys and lowlands only of this central province might describe it as an agglomerate of hills, ridges, and mountains, irregularly distributed and without system; but from a higher altitude the tops of the hills and ridges appear to mark a gently undulating plain. Viewed from about the level of this plain, the drainage channels are almost entirely hidden, and the upland surface sweeps off to the horizon, broken only here and there by peaks or mountain masses, which, rising above the general level, relieve the otherwise monotonous landscape, and indicate more resistant formations which have withstood erosion better than the surrounding terranes.

Toward the north the land is deeply covered with products of weathering, while farther south, the results of glaciation are more in evidence, the valleys being filled with glacial deposits and the rocks of the hillsides being scoured and grooved. All signs of glacial action disappear toward the northern end of the district described in this report.

Local.—The striking features of the district are the wide, prominent valleys of the Lewes river and its tributary the Nordenskiöld. These rivers have deeply dissected the region, leaving between them a high rolling upland. West of the Nordenskiöld, the country gradually becomes higher and more rugged, until eventually the mountains of the Pacific Coast system are reached. All the more important valleys of the district, as well as some of the smaller ones, are generally picturesquely spotted with lakes and ponds of all shapes. These are mostly small, but, in some cases, are several miles in length, and are generally of glacial origin. A large number of the higher valleys are floored, in many places, with muskegs that are impassable, except when frozen.

DETAILED ACCOUNT.

Lewes river, and Lake L. Serge.—From Whitehorse to Lake Laberge, the Lewes river flows in a general northerly direction through a wide valley bordered on the east by a very striking range of white limestone hills that continue north for a considerable distance past the northern end of the lake. The river is very swift for 2 miles below the Whitehorse rapids, and, to within 10 miles from Upper Laberge the current will probably average 4 miles an hour. Below this, to the lake, it is rather slack, and the bed and banks of the river are clayey and sandy, while above, where the water is more rapid, they are chiefly of gravel.

Thirteen miles below the Whitehorse rapids, the Lewes is joined by the Takhini river which flows from the west in a direction about at right angles to that of the main river. Dr. Dawson[1] has calculated that at 200 yards from the Lewes, where the Takhini has attained its full size, it is, at average low water in summer, approximately 237 feet wide, with a depth of 10 feet for one-third of its width, and with a current of 2 miles an hour; thus making the discharge about 3,600 cubic feet per second, or one-half that of the Lewes above the confluence.

The valley of the Lewes, at the head of the lake, is occupied by low swampy flats and terraces, composed—where cut by the river—of fine, often iron-stained, stratified sands, overlying glacial silts. They are, therefore, valley deposits of post-glacial age.

---

[1] G. M. Dawson. Annual Report, Geological Survey, 1887-8, Part B.

PLATE II.

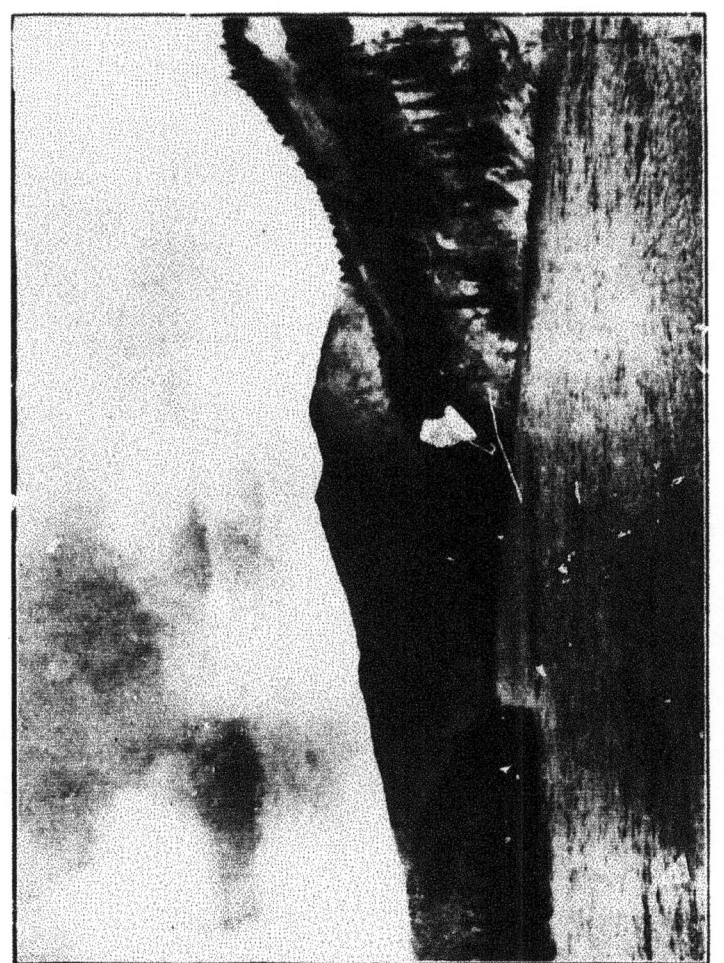

Looking down the Lewes river toward the Semenof hills.

Lake Laberge, through which the Lewes river flows, and which has a north-northwesterly trend, is 31 miles long, the upper 16 miles to below Richthofen island averaging 5 miles, and the rest probably 2 miles, in width. The lake is 2,050 feet above sea-level, is rather irregular in outline, and is bordered by a hilly or mountainous country. The white limestone mountains along the southern portion of the east side of the lake attain elevations of about 2,000 feet, at a distance of 2 or 3 miles from the shore; but toward the lower end, the hills, which rise abruptly from the valley, are only about 400 to 1,200 feet high. On the west the lake is bordered by gently sloping hills which attain heights of 2,000 feet some miles inland, and are nearly all wooded, presenting quite a contrast to the white treeless hills facing them on the east. Terraces were noted along the lake at various elevations, up to 350 feet above the water.

Two important valleys join that of Lake Laberge—Richthofen valley and the Ogilvie valley. The former trends northwesterly from west of Richthofen island, while the latter extends from the southwestern corner of the lake in a direction approximately parallel to the first. It has been suggested that the Lewes river formerly flowed through the Ogilvie valley, which is a continuation of that occupied by the lake, and which channel has been filled with morainic and other drift material during the glacial period, forcing the river to take its present course. Numerous terraces fringing the valley were noted up to 200 feet above the level of the lake.

The river at the lower end of the lake turns to the northeast, and breaks through the hills on that side, through an incision which, although existing in glacial times, possessed, in all probability, much less importance then than at present. A government telegraph office is situated on the right limit of the Lewes just below Lake Laberge.

From Lake Laberge the Lewes, to its confluence with the Teslin at Hootalinkwa—where there is also a government telegraph office—is generally known as Thirtymile river, the distance being about 30 miles. It has a general trend somewhat east of north, is very tortuous, and does not follow any particular or well marked valley; but merely a depression among irregular, lumpy, hills, seldom over 1,000 feet above the river. The valley near the mouth of the Teslin is more constricted than usual, and in this respect is remarkable,

since it is the point of confluence of such important rivers. The
valley of the Teslin appears to be the upward continuation of the
valley of the combined streams below their junction, the Lewes
flowing through a gap between high hills nearly at right angles to
the lower course of the river. Dr. Dawson[1] has measured the two
streams above the junction and gives the following:—

'The Lewes showed evidence of having risen about a foot above
its lowest summer level, while the Teslin was probably near its
lowest summer stage. If we subtract the volume of water repre-
sented by this extra foot in depth, the discharge of the Lewes at
summer low-water stage may be approximately stated at 15,600
cubic feet per second.

	Lewes.	Teslin.
Mean width	420 ft.	575 ft.
Maximum depth (near left bank)	12 ft. (near right bank.)	18'-4"
Sectional area	3,015 sq. ft.	3,809 sq. ft.
Maximum velocity	5.68 miles per hr.	2.88 miles per hr.
Discharge per second	18,664 cub. ft	11,436 cub. ft.

The actual width of the Lewes at a point 9 miles below the
mouth of the Teslin is 480 feet, and the current there is 4.84 miles
per hour. This is taken as being representative of the stream.

An evident expression of the uplifted peneplain occurs to the east
of Hootalinkwa, and is particularly emphasized in the vicinity of
Livingstone creek, a tributary of the Big Salmon river. There,
from the summits, the upraised and subsequently dissected base-
level surfaces are very apparent, with small rounded hills and ridges
rising above the surrounding upland, and representing the only
remaining elevations in the original topography.

For 3 or 4 miles below Lake Laberge the current of Twentymile
river is rather slack, below that, however, it averages probably 6 miles
an hour.

From the mouth of the Teslin to that of the Big Salmon the
Lewes has a course due north. The hills, bordering the valley near
Hootalinkwa, rise to between 1,000 and 1,500 feet above the river;
but gradually decrease, in a few miles, to 800 or 900 feet, at which
elevation they continue to near the Semenof hills, through which

[1] Annual Report of the Geological Survey, 1887-8, p. 153 B.

PLATE III.

Kynocks roadhouse on the Whitehorse-Dawson wagon road, 62½ miles from Whitehorse.

both river and valley are exceptionally constricted. This range of hills, which is dissected by the Lewes some 5 miles above Big Salmon, is about 5 miles wide, has a general northwesterly trend, and consists of rounded and wooded hills rising to heights of from 1,500 to 2,000 feet above the river. A short distance above the mouth of the Big Salmon river, and on the left limit of the Lewes, is the government telegraph office, generally known as Big Salmon.

Below Big Salmon river, the Lewes turns to the west, almost at right angles to its previous course, and flows in a northwesterly direction to Tantalus. From thence to Rink rapids it trends in a general way almost due north, the average current from Big Salmon being about 4 miles an hour.

Immediately below Little Salmon, to near the Rink rapids, the river meanders greatly, but above this—from Big Salmon to Little Salmon—it is not nearly so tortuous, and for a considerable portion of the distance is not so swift as in most places. For 8 or 10 miles midway in this latter stretch, both river and valley are constricted, but just before reaching Little Salmon, the Lewes comes into a wide flat basin extending several miles from the river on its left limit. Below this, the valley narrows again, and, to Five Fingers, is perhaps narrower than the average. Both Big Salmon and Little Salmon rivers, for several miles from the Lewes, occupy broad, flat, depressions.

The hills on the right bank of the Lewes, in the vicinity of Little Salmon, are open, high, and bare, and attain heights of from 1,000 to 1,500 feet, but below there to Five Fingers, they seldom exceed 900 to 1,000 feet above the river in the vicinity. Terraces rising to 200 feet above the river were seen in many places, extending back occasionally at about that level to the base of the hills.

Dr. Dawson[1] has estimated, at the end of August when the water in these rivers is at its lowest summer level, that the Nordenskiöld river, 200 yards from where it enters the Lewes, is 80 feet wide, with an average depth of 6 inches; that Little Salmon river is 100 feet wide with an average depth of 3 feet; and that the Big Salmon river, near where it joins the Lewes, carries 2,736 cubic feet per second.

Braeburn-Kynocks Coal Area.—This area is named after the two road-houses which are situated within its boundaries on the

[1] G. M. Dawson. Annual Report, Geological Survey, 1887-8, Part B.

Whitehorse-Dawson road, and is 35 miles long from east to west, with an average width of 12 to 14 miles. It includes on the east the north end of Lake Laberge, and extends to the west across both Klusha creek and Schwatka river, the East and Middle branches respectively of the Nordenskiöld river. The distance from the western edge of Lake Laberge in a line almost due west to Klusha creek at its nearest point, just south of Lake Braeburn, is 18 miles; and from here due west to the Schwatka river is 9 miles. However, 5 or 6 miles farther south, these two branches come within 4 miles of one another, Klusha creek, along which the Whitehorse-Dawson road has been built, having made a wide curve to the west.

Lake Laberge is bounded on the west by fairly abrupt walls rising to the level of the plateau, from the edge of which the surface gently rises until, at a distance of 7½ miles from the lake, a low divide, extending along near the centre of this upland interstream-area, is reached, which is 1,700 feet above the lake—the highest summits here being over 4,300 feet above sea-level, or 2,250 feet above Lake Laberge. From here continuing westward the country gradually descends again to near Klusha creek, where there is a sharp drop to the valley which is here between 2,350 and 2,400 feet above the level of the sea. Between this valley and that of the Schwatka river—which has an elevation of between 2,100 and 2,200 feet—is another range of hills, or rather a narrow but elevated belt. the highest eminences of which are between 3,700 and 3,800 feet. The highest summit on the map is that of Vowel mountain, 3,133 feet high, and lying about 2 miles west of Schwatka river.

The Braeburn-Kynocks area is really a narrow strip of upland, through which three, deep, parallel incisions have been made in a northerly direction, one near the eastern, and the other two near the western edge. There are also a number of tributary or transverse valleys, the most important, probably, being the Ogilvie valley, which trends, as described above, in a northwesterly direction from the northwest corner of Lake Laberge. One important transverse valley extends from Klusha creek to Schwatka river, between Division and Cub mountains. Others, often deeply dissecting the district and constituting the minor drainage ways of the district, are shown on the map.

The valleys generally contain numerous lakes of various shapes, and up to several miles in length. Some of these are connected by

PLATE IV.

Packing in the Nordenskiold valley.

streams with each other, but others have, apparently, no inlet or outlet. Muskegs are very common, especially in the higher valleys.

Numerous terraces, from 10 feet to several hundred feet above the valleys, extend along both Klusha creek and Schwatka river.

With the exception of the highest ridges and summits, and a few south slopes, the country is mostly wooded. About one-third of the forest is white spruce, the remaining two-thirds being chiefly composed of balsam poplar, aspen, willows, and a small amount of black pine. Scrubby spruce extends up to an elevation of 4,000 feet. Ridges and hills above this, and up to 4,300 feet, are chiefly covered with a thick growth of dwarf birch, and frequently some willows. Above 4,300 feet, as on Vowel mountain, only moss grows.

Tantalus Coal Area.—The portion of this area along the Lewes river has been described above, and the topography to the north of the river is quite similar to that of the Braeburn-Kynocks area. In this case, however, the area surveyed lies parallel to the Nordenskiöld river, which, with the first range of hills to the west, continues along the western edge of the map. The valley of the Nordenskiöld is quite flat, 1 to 1½ miles wide, and has, toward the north, quite abrupt walls—the slopes farther south being, however, more gentle. The higher eminences on each side attain altitudes of between 4,000 and 4,300 feet.

The valley bottom has been filled to considerable depths with glacial sands, gravels, etc., and the river is extremely tortuous—the slightest obstacle causing it to deviate from its course, as a channel is easily made in the loose material, through which the river has not yet had time to wear its way to bed-rock. The valley is also picturesquely dotted everywhere with lakes, often very beautiful, and of almost every shape and size, some small ponds, others several miles in length; and they are frequently unconnected with each other, or with the river that often winds around amongst them, sometimes even within a few yards, without touching or influencing them. They represent the positions of the last remnants of the retreating glaciers, and sufficient time has not yet elapsed for them to have become drained or filled.

The northern continuation of the Ogilvie valley, from the northwest corner of Lake Laberge, joins that of the Nordenskiöld river 6 miles below Montague, and is here, as to the south, a very broad, important depression containing a large number of lakes, and in it

is the old winter road from Lake Laberge to Montague, but which
now has not been used for several years. Near its northern end,
the valley is joined by a branch valley from a northeasterly direction, in which is a particularly long and beautiful chain of lakes
which are drained through Mandonna creek, a stream about 4
miles long, which enters the Lewes river from the south, nearly
opposite Eagles Nest, 5 miles below Little Salmon. The most
southerly of these lakes—Frank lake, trends N 20° W, and is over
5 miles long, with an average width of about 1 mile.

Climate.

Of late years, since the construction of the White Pass and
Yukon railways, and the building of lines of boats on some of the
most navigable waterways, the former popular impression that the
Yukon Territory was a region almost impossible of access, and
covered with perpetual snow and ice, has been completely dispelled.
The climate of the southern Yukon, especially that of the portion
described in this report, is similar to that of many districts in
British Columbia, and other prosperous northerly mining camps
of the world, where, in actual mining, the difficulties are not more
formidable than in localities farther south. All the necessary out-
side and surface work in connexion with mining and similar indus-
tries may be continued at least six months in the year. Moreover,
on account of very long days in this northern latitude, surface work
may be performed during a considerable part of the summer by
night as well as by day without the aid of artificial light. The
ground, in many places, is continuously frozen to varying depths,
but this does not interfere with mining, except while at, or near,
the surface.

Altogether, considering the long days and the almost constant
sunshine, the summer months in this region are particularly pleasant
and delightful. The rivers generally open early in May, but the ice
remains longer on the lakes; that on Lake Laberge generally block-
ing navigation on the Lewes river until the first week in June.
Slack water stretches freeze over shortly after the middle of October;
but during some seasons the rivers remain open until well on in
November.

Agriculture.

The wide, flat, and extensive valleys of the Nordenskiöld and its
branches are in many places covered with very luxuriant growths

et wild grasses. Timothy, or oats, where they have been accidentally scattered along the roads, have been observed to grow well. Many varieties of vegetables grown in Dawson, Whitehorse, and at intermediate points, compare very favourably with those imported. Moreover, it is well known that horses generally winter safely in these valleys without being fed. Further, during the season of 1905, the writer made a collection of the characteristic plants of the district (see list below), which has been examined by Professor John Macoun, botanist of the Geological Survey, who considers they indicate that the district is suitable for agriculture, the specimens being similar to those found in the Canadian Northwest, particularly in the vicinities of Prince Albert, and Edmonton. For these and many other reasons, it is considered that the more favourably situated parts of this district are suitable for stock-raising, and for agricultural purposes.

Fauna and Flora.

The following are the principal varieties of trees and shrubs in the district:—

Picea alba White spruce.
Pinus Murrayana—Black pine.
Populus tremuloides Aspen poplar.
Populus trichocarpa—Western balsam poplar.
Betula tluskana White birch.
Betula glandulosa Dwarf birch.
Salix Brownii Arctic willow.
Salix cordata.
Rosa acicularis.
Alnus fruticosa.
Shepherdia Canadensis.
Viburnum pauciflorum - High-bush cranberry.
Ribes Hudsonianum—Black currant.
Ribes rubrum—Red currant.
Vaccinium—Blueberry.
Amelanchier florida—Saskatoon berry.
Ledum Groenlandicum—Labrador tea.

About a third of the whole district is wooded with white spruce, which is, in places, as much as 2 feet in diameter, 12 inches to 18 inches being, however, more common. These trees are tall and well-grown in the valleys and on the hillsides, and extend in shrubby form to an elevation of 4,000 feet above sea-level, above which altitude trees do not grow in this district.

Small groves of black pine were seen in a few places in the river valleys, and a few were noted on the hillsides.

Aspen poplar, balsam poplar, and willows, extend over about one-half of the country. They grow in the valleys and on the hillsides; and the willows are even found interspersed with the dwarf birch on the higher ridges up to an altitude of 4,300 feet, above which only mosses grow.

Practically the whole district, except the limestone hills, the highest summits, and a few south slopes, has a forest covering of some description.

Wild fruits were found in most of the localities. Mossberries, blueberries, and high-bush cranberries are very plentiful; while strawberries, red and black currants, and Saskatoon berries, were also found.

The following is a list of the characteristic plants of the district, the grasses excepted. These were collected by the writer during the season of 1908, and have been named by Professor J. Macoun :—

> *Primula farinosa*—Bird's-eye primrose.
> *Mertensiana paniculata*—Lungwort.
> *Polemonium pulchellum.*
> *Arnica alpina*—Arnica.
> *Pyrola grandiflora*—Small shin leaf.
> *Potentilla fruticosa*—Shrubby Cinquefoil.
> *Potentilla Anserina.*--Silver weed.
> *Arctostaphylos Ura-ursi.*—Bearberry.
> *Linum perenne*—Flax
> *Anemone Nuttalliana*—Open wind-flower.
> *Hedysarum MacKenzei.*
> *Lupinus arcticus*—Northern Lupine.
> *Astragalus hypoglottis.*
> *Corydalis aurea*--Golden corydalis.
> *Aster alpinus*—Alpine aster.
> *Potentilla nivea*—Northern Cinquefoil.
> *Anemone Richardsonii*—Golden anemone.
> *Saxifraga tricuspidata*—Saxifrage.
> *Fragaria cuneifolia*—Strawberry.

Moose and caribou are the only species of deer in the district. The former are fairly plentiful in many localities, but the latter are rather scarce. Black and grizzly bears, wolves, wolverine, beaver, martin, and otter, are somewhat common, and lynx are very plentiful. Cross, black, and silver foxes are also seen occasionally. Rabbits, and grouse of different sorts, which a few years ago were very abundant, are now rarely seen.

Salmon go almost to the heads of the streams tributary to the Lewes river, and the Indians, who generally catch them by building fish-traps across the small streams, depend on them, to a great extent for food. The rivers, streams, and lakes of the district are generally well supplied with fish, chiefly:—

Oncorhynchus chiucha—King salmon.
Coregonus Nelsoni—White fish.
Salvelinus Namaycush—Lake trout.
Thymallus signifer—Grayling.
Esox lucius—Pike.

GENERAL GEOLOGY.

General Statement.

Regional. Looking toward the west from any of the higher summits of the district described in this report, the high, rugged, generally snow-capped, ranges of the Pacific Coast system can be seen extending in a northwesterly direction. To the east of this mountain province stretches the wide, rolling, somewhat mountainous, interior Yukon plateau, chiefly underlaid so far as is known, by schists, gneisses, cherts, slates, limestones, conglomerates, sandstones, shales, dacites, andesites, basalts, breccias, and tuffs, generally covered by a thin mantle of Quaternary deposits.

The greater part of this interior country is thought to be floored by pre-Ordovician schists and gneisses, ... ain in places by Devono-Carboniferous (?) cherts and slates. . . . wever, only small outcrops of these older rocks were seen in or near the district here considered. Overlying them are limestones, forming the white, generally treeless ranges which, from their striking contrast to the rest of the landscapes can be readily distinguished for long distances. Newer than the limestones, and covering them over wide areas, are great thicknesses of Jurasso-Cretaceous sediments, in turn cut, broken, and covered by an extensive series of andesites, basalts, etc., which have subsequently been covered by another series of basalt, presumably of Tertiary a . . . .

Toward the southern part of the Yukon, where glaciation has played an important rôle in shaping the topography, the valleys are filled with heavy deposits of glacial gravels, clays, sands, and silts, which extend some distance up the hillsides—traces being even found on some of the higher summits. Evidence of the action of the northwesterly moving ice gradually disappears toward the north; until north of Tantalus no distinct markings are seen; glacial silts are found, however, much farther down the river. Beyond this line of the former glaciers a large portion of the country, often including the tops of ranges and mountains, is covered with great thicknesses of recent material, either formed directly by the weathering of underlying rocks, or transported by the present streams and rivers.

PLATE V.

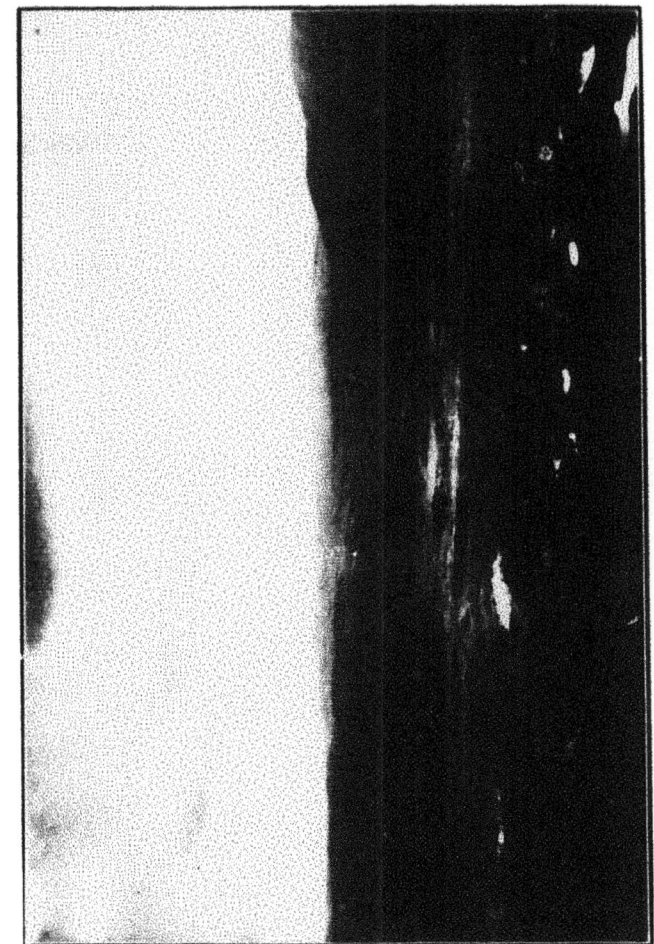

Looking across the Norddiskrold Valley from Borchie mountain. The valley is dotted with numerous lakes of glacial origin.

Local.—Exposed only on one small group of hills, situated on the western edge of the Tantalus coal area, are light-coloured, generally reddish, mica gneisses, and fine-grained, dark bluish to greenish, schistose amphibolites. These rocks, comprising the Razor Mountain group, have for reasons given below, been considered to be probably pre-Ordovician. It is also thought that the greater part, if not all, of this district is underlain by pre-Ordovician gneisses and schists now hidden by younger formations.

Two or 3 miles south of the area of the Razor Mountain rocks is a small exposure of cherts and slates of the Montana group, lithologically very similar to members of the Lower Cache Creek series in the Windy Arm district.

The oldest rocks exposed at all extensively in this district are the Braeburn limestones, which have a wide development. These are over 2,000 feet thick—possibly several times this thickness—and are generally semi-crystalline to crystalline, and white, to light blue, in colour.

Invading the limestones and in places overlying them, are the Nordenskiöld dacites, generally reddish to reddish-blue in colour. These rocks are probably not very widespread; for they have been noted in the Yukon Territory, only along the Nordenskiöld river.

In most places—superimposed upon the limestones, or upon the dacites if present—occur the Lake Laberge series, estimated at 3,800 feet thick, and composed of conglomerates, sandstones, greywackes, slates, etc. These beds cover a large portion of the district, and are overlain conformably by the Tantalus conglomerate, which, where not eroded, is at least 1,000 feet thick; the pebbles composing it consisting of cherts, cherty quartzite, or slate. The two coal horizons of this district occur, respectively, near the top of, and at 12 to 300 feet below these conglomerate beds. The Laberge series and Tantalus conglomerates are both of Jurasso-Cretaceous age.

Cutting, breaking, and overflowing these sedimentaries in many places are the Schwatka andesites, and the andesites, basalts, tuffs, etc., of the Hutshi group, the latter being contemporaneous in age with the Schwatka volcanics. The rocks of the Schwatka series are light coloured, greyish, reddish, bluish, etc., with generally large, easily discernible, phenocrysts of feldspar, mica and hornblende. The members of the Hutshi group are generally dense, and dark-greenish in colour. Included in this series, near the top, are beds of dark, fine-grained, greywacke shales so closely associated with the andesites, etc., that, except in very detailed work, it would be impossible to map them separately.

Newer than all the above-mentioned formations are the Klusha intrusives, chiefly syenite porphyries, developed in a few places only, generally in the form of small stocks or dikes.

The most recent consolidated terrane in the district is the Carmack group, consisting of basalts and basalt tuffs, extensively developed in the northern portion of the Tantalus area.

All the main valleys are filled, to considerable depths, with glacial drift, that, toward the south, extends up the hillsides as well. Overlying this everywhere is a thin mantle of recent material composed of sands, gravels, silts, etc., laid down by the present waterways, and also peat, muck, soil, etc. A noticeable feature of the country is a very even layer of volcanic ash, 6 to 13 inches thick, which covers the entire region. This ash is newer than the silts—the most recent of the glacial products—and is so near the surface that the grass roots extend down into it.

TABLE OF FORMATIONS.

System.	Formation.	Lithological Character.
Quaternary		Gravels, sands, silts, clays, muck, so volcanic ash, etc.
Tertiary?	Carmack basalt	Basalts and basalt tuffs.
	Klusha intrusives	Chiefly syenite porphyry.
	Schwatka andesites	Mica and hornblende andesites.
	Hutshi group	Augite andesite, basalts, tuffs, shales, etc.
Jurasso-Cretaceous.	Tantalus conglomerates	Conglomerates, sandstones, etc., composed chiefly of cherts, slates, and quartzites. Coal bearing.
	Laberge series	Conglomerates, greywackes, sandstones, shales, etc. Upper beds, coal bearing.
	Nordenskiöld dacites	Dacites, generally reddish-blue in colour.
Carboniferous?	Bradburn limestones	Limestones, generally semi-crystalline to crystalline, in places siliceous.
Devono-Carboniferous?	Montague group	Cherts, cherty quartzites, slates, and argillites.
Pre-Ordovician.	Razor Mountain group	Gneisses and schists.

PLATE VI.

Looking down the Nordenskiold from Breder mountain, seven miles above Tantalus. The valley here, as in most places, contains numerous lakes, generally unconnected with the river.

Descriptions of Formations.

Razor Mountain Group.—The gneisses and schistose rocks—assigned to the Razor Mountain group—are exposed to the south of Razor mountain, on the western edge of the Tantalus coal area. They cover an area about 3 miles long, from north to south, and extend at least 1 mile to the west, but how much farther was not determined, this being the western edge of the area mapped.

The members of this group are chiefly light coloured, generally reddish, mica gneiss, and fine-grained, dark, bluish to greenish, schistose amphibolite. Under the microscope the gneiss is seen to contain a large proportion of quartz as well as some orthoclase, microcline, oligoclase, biotite, and subordinate muscovite, the minerals showing wavy extinction, the result of pressure. The amphibolite, a schistose, diopside-bearing, amphibolite, consists chiefly of diopside and plagioclase with also some larger phenocrysts of common green hornblende, and a small amount of biotite.

These rocks lithologically much resemble some of the 'Older Schistose rocks,' to the north, in the Klondike district and in Alaska, considered by R. G. McConnell[1], A. H. Brooks[2], and others, to be pre-Ordovician. Because of this lithological similarity, the rocks of the Razor Mountain group have also been assigned to the pre-Ordovician.

In the Conrad Mining district, to the south, similar schistose rocks and gneisses occur which also are probably pre-Ordovician[3]. The occurrence of these isolated exposures of similar rocks over such an extensive area probably indicates that a great part of the region is underlaid by them.

Montague Group.—The members of this group are found, in the district mapped, only in one locality: situated on the north side of the creek between Jimmie and Poplar mountains, the extent of the exposure being 1,500 feet by about 1,000 feet. An outcrop of similar rocks was noted on the Lewes river opposite the mouth of Fife creek, a few miles above Big Salmon.

[1] R. G. McConnell.—Report on the Klondike district. Geol. Survey of Canada.
[2] A. H. Brooks.—The Geography and Geology of Alaska. U. S. Geol. Survey. Prof. Paper No. 45.
[3] D. D. Cairnes.—Report on Portions of the Conrad and Whitehorse Mining districts, Yukon Territory. Geo. Survey Branch, Dept of Mines. Canada.

This group consists chiefly of white, grey, bluish, reddish, to nearly black cherts, cherty quartzites, slates, and argillites. The rocks are commonly thin-bedded, the chert and slate beds generally not exceeding 1 or 2 inches in thickness. All are much folded, strained, and distorted.

As these rocks, lithologically, very closely resemble those of the Lower Cache Creek series in the Windy Arm district, which are considered to be Devono-Carboniferous, it is thought that the Montague group is probably of the same age. However, all that is definitely known, is that they are older than the surrounding members of the Hurshi group.

Braeburn Limestones.—These limestones extend along nearly the whole of the east side of Lake Laberge, and occur in a number of places along the Lewes river between Laberge and the Teslin river. Below the Teslin the first of these beds noticed occurs on Eagles Nest, a limestone hill on the right side of the Lewes, 5 miles below Little Salmon. On the opposite side of the river, to the south of Eagles Nest, the Braeburn limestones are extensively developed, but were not seen again from that point down the river to Five Fingers.

In the Braeburn-Kynocks area, in addition to outcropping quite extensively along the north and east sides of Lake Laberge, these limestones are also exposed to the north of, and form the eastern part of Surprise mountain, 4 miles east of Lake Laberge. They are also exposed for 18 or 20 miles along the eastern edge of the southern portion of the Tantalus area, forming a part of the wide limestone belt which extends to the east around Frank lake and other lakes in the valley, and along Mandanna creek to the Lewes river, opposite Eagles Nest. The limestone area also extends westward to the east side of the Nordenskiöld valley opposite Braeburn, and continues along the valley, to the north, past mile-post 87 from Whitehorse, on the Whitehorse-Dawson road. It also crosses the valley a few miles below Braeburn, and is found for some distance farther to the west.

The limestones are generally subcrystalline, in places approaching marble, but in others are rather flaggy and argillaceous or even siliceous. They are commonly white to light bluish in colour, and exposures can generally be distinguished at some distance, since, in this district, the limestone hills are usually bare of vegetation. This

Plate VII.

Looking across the Nordenskiöld river from opposite Montague, showing the wide flat valley, and the plateau nature of the country to the east.

terrace, therefore, can generally be mapped more definitely than others in the same locality, as everywhere, except in the valleys, the rocks are exposed to view.

This limestone appears to be the same as that of the upper Cache Creek series, found to the south, in the Conrad Mining district, and therefore, of Carboniferous age. The evidence, however, afforded by the fossils collected has been of an indefinite character, and the inference that the two formations are of the same age has been made mainly on lithological grounds. In both districts there are several thousand feet of these massive limestone beds, and they have been traced almost continuously from Windy Arm to Lake Laberge. Still, until more fossils are found, or more proof obtained to further substantiate the correlation, the name Braeburn limestones will be retained, and the group will be only provisionally placed under the Carboniferous.

Nordenskiöld Dacites.—These dacites were found only at a few places in the Tantalus area, chiefly along the eastern edge of the Nordenskiöld valley, where a narrow band, generally only a few hundred feet wide, extends along the base of the hills from the foot of Montague mountain to about 3 miles south of the Montague road-house. They apparently cross the valley at this point; since they outcrop and are again seen on the opposite side, forming the main part of the hills on Khusha creek opposite Conglomerate mountain, at the extreme south end of the Tantalus coal area.

The Nordenskiöld dacites are generally coarse-textured, and, on fractured surfaces, reddish-blue to reddish-grey in colour, the feldspars, biotite, and quartz being in most cases quite easily discernible. On account of the decomposition, chiefly of the plagioclase and iron-bearing minerals, the rocks, in many places, readily weather to a reddish, crumbly mass, and are generally red on weathered surfaces.

Under the microscope the rock varies from a finely, holocrystalline ground-mass to one consisting mostly of glass, and is composed chiefly of quartz and feldspar. The phenocrysts are chiefly quartz, bytownite, biotite, and accessory magnetite. The quartz constitutes a great portion of the rock.

These dacites invade and overlie the Braeburn limestones, and are overlain by the members of the Laberge series.

Laberge Series.- The development of this series is more extensive than that of any other terrane in the areas examined. An extensive development of it extends from Lake Laberge due west to
Klusha creek, and from Braeburn lake, in the Klusha Creek valley,
west to the Schwatka river. Its members are also found over the
greater part of the southern half of the Tantalus coal area west of
the Nordensköld river, where the outcrops extend to the valley bottom, opposite the mouth of Klusha creek, and thence continue along
the valley, to the south, to opposite the 87 mile-post from Whitehorse, where the underlying limestones come to the surface. To the
south and west of this the Laberge beds are covered by recent
volcanics. They are also developed along the west side of Lake Laberge, and in a few places on the east side also along a considerable
portion of the shore between Laberge and Hootalinkwa, and thence,
occasionally, to Little Salmon. From Eagles Nest, they are found
continuously along the right bank of the river to Tantalus butte,
and from a short distance below this place there are almost con-
tinuous exposures to the Five Fingers rapids; which are caused by
the basal, coarse conglomerate beds of the series. On the left side
of the Lewes, from 10 miles below Eagles Nest, the Laberge beds
outcrop most of the way to Tantalus, where the overlying Tantalus
conglomerate commences.

A complete section of the Laberge series was not seen in any
one place where the thickness could be accurately measured. However, parts of the section were examined in several places, and the
total thickness estimated to be at least ,800 feet.

Heavy beds of coarse conglomerate, 600 to 700 feet thick, occur
at, or near, the base of the series. But other strata, in places, lie
between it and the underlying series—limestones or dacites, as the
case may be. On Sunday mountain, at the southwest corner of Lake
Laberge, the conglomerates are underlain by 20 feet of alternating
clays and coarse-grained and light-coloured sandstones, the clays in
beds 12 inches thick, and the sandstones in strata 2 inches to 3 inches
thick. The beds intervening between these and the underlying limestones are concealed. At the Five Fingers rapids about 100 feet of
fine dark shales are exposed below the conglomerate. But in most places
where a contact was seen between the members of the Laberge series
and the limestones, the basal conglomerates directly overlie the latter,
there being in all cases an unconformity between the Laberge rocks

and the limestone. To the east of Last mountain, in the Tantalus area, the coarse conglomerate was seen distinctly overlying the limestone; the lower beds of the former containing more and more pebbles of the latter, until, at the contact, the conglomerate was practically all limestone. Similar occurrences were noted on the east side of Lewes mountain and in other places. However, on the west side of Conglomerate mountain, where the Laberge series overlies the Nordenskiöld dacites, the amount of dacite gradually increases in the lower conglomerate, so that it is difficult to determine where the conglomerate ends and arkose commences, and where arkoses give place to true dacites. The graduation from dacite to conglomerate is also quite apparent along the foot of the hills skirting the east side of the Nordenskiöld valley for several miles on each side of Montague road-house.

Granite pebbles are the most numerous in this basal conglomerate; but there are also a great number of fine-grained greenish volcanics and greywackes. Limestone and dacite pebbles are not very plentiful except near contacts with these rocks. The conglomerate is typically coarse, the boulders being often 18 inches diameter, but generally much smaller.

The members of the lower half of the Laberge series are exposed just below Eagles Nest. ... , heavy-bedded coarse conglomerates, 600 to 700 feet thick, over... the Braeburn limestones unconformably, and are overlain by 50 feet of coarse, almost white, calcareous, loosely cemented sandstone[1], showing little evidence of bedding. Superimposed upon this is a series about 650 feet thick, the rocks of which are only partly exposed. They appeared to be chiefly light greyish brown, and greenish blue greywackes[2], with some interbedded dark shales. Fossil tree trunks were quite plentiful in the softer, thickly-bedded sandstones. Over these are 500 feet of dark, almost black, thinly-bedded shales, interbedded with which are a few hard, coarse, generally dark, greywackes. Some of the shales are carbonaceous, and contain particles and seams of

[1] In this report a distinction is made between the terms 'sandstone' and 'greywacke.' The former includes sedimentaries, consisting chiefly of small quartz grains cemented with some sort of binding material; while greywacke is used for those similarly textured, consolidated sediments, that consist chiefly of a widely varying mixture of quartz, orthoclase, and plagioclase grains, as well as rounded or sharp cornered fragments of quartzite, schist, diabase, granite, and other rocks.

[2] Ibid

... up to 1 foot thick. Above ... ... four feet of reddish calcareous sandstones and conglomerates, in which are also seen thin
... ... none being noted, however, very much thick ...

From Eagles Nest, down the river for several miles, the strike
of the rocks is about parallel to the general direction of the valley
so that the beds continue along the river bank, at about the same
elevation as Extremity and D... ... mountains ... the Tantalus area.
For ... here the beds gradually dip toward the southwest, exposing the
overlying reddish, greenish and almost white, coarse sandstones
which ... in turn overl... ... the Tantalus conglomerates.

An almost complete section of the upper members of the Laberge
... is exposed, ... creek, at the foot of ... ... to the ... of
the main mountain, ... Klusha creek and Schwatka river. Here
the ... the sandstones and conglomerates similar to those at Eagles
Nest ... about 200 feet thick. The reddish colour is due to iron
exhibit... the calcareous cementing materials, the boulders and pebbles consist chiefly of a ... ... and fine-grained greenish ... . Over
these red beds are seen to come ... chiefly ... coarse, very generally
coarse-... ... beds ... sandstones, interstrat... ... ...
... some dark yellow sandstone, the colour of the greenish ...
being due chiefly to the chloritic cementing material. Above these
are 1000 feet of almost white, coarse sandstones composed almost
entirely of clear quartz pebbles and a white cement. Occasionally
the ... material contains some iron oxide, causing the sandstone
t ... have a reddish, mottled appearance. Toward the top of this white
sandstone are occasional interstratified dark shale beds, in which are
the coal seams of the lower coal horizon of this district. Overlying
the white sandstone is the Tantalus conglomerate. All the outcrops
which were noted on Extremity mountain, Thuber mountain, and
... ... the river from the latter, are of this white sandstone.

The rocks along the western side of Lake Laberge are lower
members of the Laberge series, and consist chiefly of greywacke, a
few beds of conglomerate, and some dark greywacke shales. The
greywackes are generally slightly indurated, fine-grained, and greenish in colour, and are either quite massive or show more or less
bedding. Where these are fine-grained, and without marked bedding planes, they might easily be mistaken for volcanics, since for
t... thicknesses of ... or more feet they have the same massive appearance
throughout. In fact, in the field they are in places thought to be

tuffs, but under the microscope they proved to be greywackes, containing in some cases considerable volcanic material. The dark shales, which are associated with these greywackes, are in places several hundred feet thick, generally thinly-bedded, nearly black in colour, often quite hard, and are in places true slates with induced secondary cleavage. Some of the beds are heavily iron-stained from the oxidation of pyrite.

Good sections of parts of the Laberge series were also seen on Joe and Fossil creeks between Belleview and St. Hilary mountains, in the western portion of the Braeburn-Kynocks area. Up Fossil creek, dark, almost black, thinly-bedded, often iron-stained shales outcrop for a mile or more, the strike of the beds, which are here about 500 feet thick, being almost parallel to the direction of the creek. These shales are similar to those along Lake Laberge, just below Eagles Nest, and elsewhere, and are almost certainly of the same horizon. They also much resemble the Benton shales of the foothills of the Rockies in Alberta. On Joe creek, about 2 miles to the southeast of the shales on Fossil creek, there is an almost continuous rock exposure for nearly 2 miles, the beds being chiefly coarse, dark greenish, greywackes; hard, brown, brownish grey, grey, and even dark red, greywackes; a few softer beds of yellowish and greyish sandstones; and a considerable thickness of dark shales similar to those seen on Joe creek. Remains of tree trunks are very plentiful in these strata.

The section seen on Joe creek is composed of strata of the same horizon as the beds on Bunker hill, Belleview mountain, St. Hilary mountain, Black Jack, South Extension, Contact mountain, between Braeburn lake and Lake Laberge, and on the northern part of the Tantalus area; but on the hills only the sandstones and harder strata outcrop, since the softer beds and shales, which weather more easily, are covered with drift.

On the west face of Montague mountain, facing the Nordenskiold river, is a practically complete section of the Laberge series from the Nordenskiold dacites, which underlie them at the base, to the Tantalus conglomerates on the top of the mountain. The beds are mostly covered with superficial deposits, and only the basal coarse conglomerate of the series and a few light-coloured grey, yellow, and light brown sandstone beds are exposed. To the south of Montague mountain, on Flower mountain, and on the hills to the west and south of the latter, the exposed members of the Laberge series are chiefly conglomerates and sandstones. The whole top of Conglomerate mountain is also covered with the typically coarse, basal conglomerate of the series.

On a small creek about 2 miles south of Coal creek, on the east side of the Nordenskiöld valley, is a good section of part of the white sandstones that underlie the Tantalus conglomerate. These are here soft, friable, coarse-grained, and weather readily into gravel; they are prevailingly white, light yellow, and greyish in colour, a few beds being, however, light brown. There are, in places, numerous intercalated, narrow, persistent, black shale beds from ½ inch to 1 inch thick, giving the whole section a finely striped appearance. In places, bands 5 to 10 feet thick occur which are nearly half of these dark shales. Just above this horizon, although not here exposed, is the lower coal horizon.

Along the Whitehorse-Dawson road, a few miles from Whitehorse, sandstones, greywackes, conglomerates, and shales probably belonging to the Laberge series, occur at a number of points. Greenish greywackes outcrop between the 6th and 7th, and 7th and 8th mile-posts; finely-bedded, dark shales are found near the 12th; and between the 11th and 16th mile-posts are frequent exposures of fine-grained, reddish, also coarser, light-coloured, greywackes, and conglomerate.

Portions, at least, of this Laberge series are of the same age as the Tutshi series of the Conrad Mining district; but the former terrane is much the thicker, including as it does considerable thicknesses of greywackes and other beds that are not found to the south. At, or near, the base of both are thick beds of similar coarse conglomerates, and both are capped by the Tantalus conglomerates. The fossils found in the Tutshi rocks are considered by Dr. J. F. Whiteaves of the Geological Survey to be lower, or lower middle, Cretaceous; and those from the Laberge series to be Jurassic or Cretaceous—the specimens being indefinite. Hence, while both series are, in a general way, believed to be of the same age, still, one may contain beds older or newer than the other. Therefore, the evidence for the correlation is stronger, and until more fossils have been found and examined, it has been decided to give the new name ' Laberge series ' to these sediments to the north.

Concerning the fossils collected from the Laberge series, from the east of Vowel mountain, and other places, Dr. Whiteaves, after provisionally naming some of them, adds: ' These fossils, unfortunately, do not give any very definite information as to the probable age of the rocks from which they were collected. Only three of them are specifically identified, with a reasonable degree of probability, viz., *Trigonia Dawsoni, Nerinœa Maudensis,* and *Rhynchonella orthidoides.* The others are either undescribed or indeterminable.

The following tentative conclusions are suggested by these fossils —:

(1.) That all the fossils are from the same geological horizon.

(2.) That they are either Jurassic or Cretaceous.

(3.) That it is perhaps most probable that the geological horizon which they indicate is the Queen Charlotte formation of Dr. G. M. Dawson, which, as Stanton says, ' includes strata belonging to both the Horsetown and the Knoxville, and probably still older beds.'[1]

Tantalus Conglomerates.—The Tantalus conglomerates, so named from the occurrence in them of the coal measures of the Tantalus mine, form the main part of Tantalus butte, and extend along the top of the ridge to the north of the butte to where, at 2½ miles from the Lewes river, they are covered by the Carmack basalts. They are also exposed along the valley for three-fourths of a mile to the southwest of, and nearly 2 miles in an easterly direction from Tantalus, and extend to the south to the first high summits about 1 mile distant. Southeast of this along the top of the ridge, for 2½ miles, the formations are concealed by superficial deposits, where the conglomerates are again exposed. To the south of this, the Carmack basalts, and, farther south still, the Schwatka andesites cover the sedimentary formations, and it is only where the volcanics have been removed by erosion, that the underlying strata can be seen.

The top of Lone Pine mountain consists of the Tantalus conglomerates, which outcrop over the west slope of the mountain to the Nordenskiöld valley; except in a few places where they are covered by recent volcanics. These conglomerates are also developed on Porter and Montague mountains, the latter of which is 17 miles

[1] Bulletin of the U. S. Geological Survey, No. 133, p. 28.

southeast from Tantalus, and it is believed that they form practically a continuous belt extending along the east side of the Nordenskiöld valley for this distance, although covered in most places by volcanics or superficial deposits. From Montague mountain, for 40 miles to the southwest toward Whitehorse, the conglomerates have been eroded away. They again appear on Corduroy and Cub mountains, in the Braeburn-Cynocks area—the geological formations between Montague and Corduroy mountains having been folded in a wide, flat, anticline, in the centre of which are hills of Braeburn limestones, overlain on each side toward the north and south respectively by members of the Laberge series.

Corduroy mountain is entirely composed of Tantalus conglomerates which are exposed along a belt 4 miles long, in a northwesterly direction, and three-quarters of a mile wide. This belt is the continuation of the conglomerate area to the northwest across Klusha Creek valley, which at Cub, End, and Contact mountains is 6 miles long, by 1 to 3 miles wide. About 2½ miles to the southwest is a parallel belt of these beds one mile wide, and at least 9 miles long, extending along Division, Red Ridge and Vowel mountains, in a northwesterly direction, to past the northwest end of the latter.

There is also a considerable development of the Tantalus conglomerates, commencing 16 miles below Hootalinkwa, and extending down stream 5 miles to below the mouth of Fife creek.

These conglomerates are, where not eroded, at least 1,000 feet thick. The pebbles or boulders, of which they are composed, are rarely more than 3 inches or 4 inches diameter, being generally of hazel nut, to walnut size, and consisting entirely of quartzitic and cherty rocks and slates. They differ from other conglomerates in the district, all of which contain granitic and other igneous material. The pebbles and boulders of the upper beds of the Tantalus conglomerates are cemented by a siliceous binder into a very hard rock which weathers with difficulty. The lower strata have a calcareous or, often argillaceous cement, causing them to weather very readily to gravel.

The conglomerates on Tantalus butte dip to the west at varying angles, forming the east limb of a synclinal fold. The corresponding western limb has been eroded away on the same side of the river, but outcrops to the south, at Tantalus, and in it are the Tantalus mine coal seams. The southern continuation of the westerly-dipping beds from Tantalus butte outcrop on a small round hill, 1½ miles southeast from Tantalus.

Volcanic rocks have broken through and overflowed the Tantalus nerates to the south of Tantalus: the strike, however, of the s noticed on the different hills, is, in most cases, quite per- indicating that th have not been broken or disturbed to any at extent, but are, in most cases, only covered.

These conglomerates, in the Braeburn-Kynocks area, ave a general northwesterly strike. The main part of the top of Cub and End Mountains ridge, represents the irregularly eroded top of an anticlinal fold. Cub mountain being on the southwest, and End mountain on the northeast limbs respectively. To the northeast is the accompanying syncline, the northeast limb of which has been largely eroded away, a portion of it forming the southern part of Contact mountain. Corduroy mountain, across Klusha creek, corresponds to End mountain, being a portion of the north limb of the anticline. Commencing on the east side of Division mountain, and extending to the west across Schwatka river, along Red ridge and Vowel mountain, is a parallel belt of these Tantalus conglomerates, about 2½ miles from that on Cub mountain. The dips here are high, the beds being bent into a sharp synclinal fold, the summits of the ridges being near the cen re of the fold.

The eastern ends of both these belts of conglomerate have been covered by members of the Hutshi group. The western end of the more northerly belt, beyond Cub and Contact mountains, has been eroded away, leaving the underlying Laberge sediments exposed. The western end of the conglomerates extending along Vowel mountain was not seen, the formation continuing some distance, at least, past this mountain.

These conglomerates also outcrop to the south of Whitehorse, therefore it is believed that they originally extended in the form of a wide belt from below Five Fingers rapids to south of Whitehorse, and that after being subjected to a period of erosion they were invaded and overflowed by volcanics, chiefly andesites and basalts. The conglomerates are exposed now only where original uneroded remnants have either not been covered with volcanics, or where these latter have been eroded away, leaving the conglomerates exposed. On Lone Pine mountain, First mountain, and Andesite mountain, and the north end of Montague mountain, the andesites are, at present, being removed by erosion from the underlying conglomerates. To the south of Montague mountain for some distance, the conglomerates, as well, have been removed, leaving the Laberge rocks exposed.

These Tantalus conglomerates are without doubt the same as those found containing the coal seams to the southwest of Whitehorse, and described in the report on the Conrad and Whitehorse Mining districts.[1] In both localities they are composed entirely of the cherty rocks and slates, and in every way are lithologically the same. From the coal seams at the Tantalus mine fossil plants were collected, and have been examined by Dr. Penhallow, who says: ' All the material appears to be the same as the specimen of *Thyrso-teres elliptica*, Fontaine, as figured by Ward in the " Status of the Mesozoic Floras of the United States," vol. XLVIII, pl. LXXI, figs. 12 and 13; and to this the present specimens are provisionally referred. It is to be observed, however, that there seems to be some question as to the correctness of Ward's reference, since the specimen cited is quite distinct from the original type of *Thyrsopteres elliptica* as described by Fontaine (in " Potomac Flora " Vol. XV, p. 133, pl. XXIV. figs. 3, 3a), and it is quite possible that further and more complete specimens may show this to be an entirely new species. A somewhat related flora was described by me in 1898 as obtained by Mr. J. B. Tyrrell from the Nordenskiöld river. All the specimens shown, however, were specimens of *Cladopliebis*, and they indicated Cretaceous age.'

' The specimens from the Tantalus mine present a flora with the same facies as those from the Nordenskiöld river, and the whole conform to the flora of Kootanie age.' (Lower Cretaceous—sometimes assigned to the Jurasso-Cretaceous period, near the close of the Jurassic and at the beginning of the Cretaceous).

Hutshi Group.—Except in one area, about three miles wide, to the south of Razor mountain, this series is developed all along the west side of the Nordenskiöld valley from Shadow mountain, south to the Hutshi river, a distance of 15 miles, and extends from the valley west to the first line of summits and possibly much farther. During the season of 1908 the writer made a trip up the Hutshi river to examine some copper deposits, and found these Hutshi rocks for a distance of at least 10 miles up the river, in

[1] See Report No. 982, Conrad and Whitehorse Mining Districts, by D. D. Cairnes, 1901.

a westerly direction. These rocks form the greater part of the high mountain to the east of and adjoining Stutzer mountain; this development of the Hutshi group being a continuation of the belt along the left limit of the Lewes river, 8 miles below Little Salmon.

In the southwest corner of the Braeburn-Cynocks area Hutshi rocks apparently overlie practically all Ottawa, Kingston, and Hull mountains, and the south part of Belleview mountain, all on the east side of Klusha creek. Across the Klusha valley they again appear on the hills to the west of Cairnes mountain, and south of Division mountain. Continuing west across the Schwatka river, these rocks are exposed along a strip averaging about 1 mile wide, and extending along the southwest side of Red Ridge mountain. They also extend along Klusha creek several miles to the south of the southern edge of the Braeburn-Cynocks area. In many localities, especially on Ottawa and Kingston mountains, outcrops are very scarce, and few exposures were seen; but all being of this series, it is considered probable that the mountains are wholly covered with these rocks.

This series also outcrops in numerous places over the sides and on the summit of Povoas mountain, which is on the east side of the north end of Lake Laberge, and apparently covers the greater part of the mountain. It is also developed along the north end of the lake, and extends about three miles to the north, toward the summit of Lewes mountain.

There is also a small exposure midway between Monson and Miller mountains, on the west side of the Lewes river, below Tantalus. Fine-grained, green volcanic rocks, apparently belonging to the Hutshi group are found in a number of places along the Lewes river between Lake Laberge and the Five Fingers rapids. The greater part of the Semenof hills, west of the Big Salmon telegraph station, appears to be covered by dark, fine-grained, andesites, which are similar to those along the Nordenskiöld. Commencing 4 miles below the mouth of the Big Salmon river and extending downstream 11 or 12 miles, rocks similar to those of the Hutshi group are exposed in numerous places. There is also a very prominent development of these rocks about 10 miles above Little Salmon on the left bank of the Lewes.

Along the Whitehorse-Dawson road, from Whitehorse to the south edge of the Braeburn-Cynocks area, rocks referable to the Hutshi series were found in several places. The rocky range of hills just south of the Takhini road-house apparently consists, nearly altogether, of these rocks. Augite andesites outcrop along the road near the 29 mile-post. Andesites, volcanic breccias, basalts, etc., were seen in several places between the 50 and 60 mile-posts, associated with outcrops of granite seen to be distinctly older than the Hutshi rocks, which cut and overlie it.

The members of the Hutshi group are chiefly augite andesites, basalts, eruptive breccias, and tuffs. The andesites and basalts are light to dark greenish in colour, and vary in texture from those in which none of the component minerals are distinguishable with the unaided eye, to those in which plagioclase and pyroxene phenocrysts are quite large and easily discernible. The rocks in most places, except on weathered surfaces, appear quite fresh and unaltered, and, particularly the finer grained varieties, are very hard and brittle. They often contain a considerable amount of iron oxide, which causes a red coloration on the weathered surfaces. The basalts, which in many places show well-developed prismatic jointing, resemble in a general way the more recent Carmack basalts, but are never so fresh-looking. In the former the vesicular cavities are mostly filled with secondary minerals, whereas in the Carmack basalts the cavities are often quite empty. In most cases the Hootchi rocks contain enough lime—due to alteration of the feldspars—to cause effervescence with acids. On Hull mountain, Kingston mountain, about midway between Monson and Miller mountains, on the west side of the Lewes river below Tantalus, and in other places, are outcrops of eruptive breccias, and tuffs, which vary in texture from quite coarse to very fine, and which, at a distance, look like conglomerates or sandstones. These probably are the result of violent volcanic eruptions in which the erupted fragmental material has fallen back, in many cases, into molten lava. The material composing the breccias is, in a general way, the same as that of the andesites and basalts, but does not possess the same homogeneity—fragments of the coarser grained rocks often lying in a fine-grained matrix. On the first low hills to the north of Lake Laberge are some basalts and basalt breccias, as well as some amygdaloidal rocks which, under the microscope, prove to be typical spilites.

On a small creek which runs along the western edge of Red ridge, Hutshi rocks are almost continuously exposed for over 2 miles.

They consist chiefly of dark-greenish—in places almost black, augite andesites, in which the chief phenocrysts are augite and plagioclase which is generally bytownite—accessory magnetite being also often quite plentiful. These are in most cases embedded in a compact ground-mass, consisting of a mixture of glass and crystals of acid plagioclase and augite, in a second generation, in differing proportions. Chlorite, chiefly delessite, which is often in spherulitic aggregates, and which is derived from the alteration of the pyroxene, constitutes a considerable portion of the rock and helps to give it the green colour. In places, the andesites have an amygdaloidal structure, the amygdules being mostly filled with delessite. The ground-mass is generally largely altered to chlorite. Interbanded with these andesites are beds from 10 to 100 feet thick, of distinctly bedded, dark, to almost black, calcareous, greywacke shales which contain thin layers of igneous, probably tuffaceous material. These sedimentary beds occur so intimately associated with the andesites, that, except in very detailed work, it would be impossible to map them separately. They have, therefore, been included in the Hutshi group; but were only noted in this one locality.

Along the west side of the Nordenskiöld valley from Shadow mountain, south to past Poplar mountain, the rocks of the Hutshi group are apparently all augite andesites or basalts. Thin sections of augite andesite from Knob mountain, examined microscopically, have generally a holocrystalline porphyritic structure, the ground-mass being completely crystalline, and consisting chiefly of fine needle-like feldspar individuals, surrounded by pyroxene and its alteration product, chlorite. Embedded in this ground-mass are large, well-defined, phenocrysts of bytownite and augite, in about equal amounts. These sections are typical of the rocks of the Hutshi group. Sections of rocks from Jimmie mountain were examined which contained no plagioclase in the first generation, the phenocrysts being chiefly augite, generally with well developed zonal structure, and accessory iron ores. All these rocks contain considerable secondary calcite, chlorite, and often epidote, as well.

The rocks of the Hutshi group were seen cutting and overlying those of the Laberge series and the Tantalus conglomerates in numerous places. They are, therefore, more recent than these. They also resemble petrographically some members of the Windy Arm series in the Conrad Mining district, and most probably are of the same period of eruption.

Schwatka Andesites.—These andesites apparently originally covered the greater part of the Tantalus coal area, to the north of Lone Pine mountain, and between the Nordenskiöld valley, on the west, and the Braeburn limestone ranges, to the east. They possibly did not extend far north of Lone Pine mountain, however, since over most of that area the Carmack basalts directly overlie the Tantalus co glomerates, only in one or two places overlying the andesites.

To the south of Lone Pino mountain, the greater part of the country is heavily covered with superficial deposits, so that it is somewhat difficult to estimate just how much of the district is covered with the Schwatka andesites. The greater part, at least, of the southwest face of Lone Pine mountain is covered by them, as well as the top of Andesite mountain; the northern portion of Montague mountain; and the greater part of Saddle, Porphyry, East, and Cone mountains. There are also smaller exposures on Tanglefoot mountain, as well as in numerous places, on, and between, the above-mentioned mountains. Farther south, in the Braeburn-Kynocks area, the southwest part of Cub mountain is composed of these Schwatka andesites, which extend across the Schwatka River valley, and outcrop in the form of a wide dike extending along the east face of Vowel mountain.

These andesites are, in most places, coarse-textured, and greyish, bluish, or reddish, in colour, and generally contain large, well-developed phenocrysts of feldspar, mica, and hornblende. Under the microscope, they are seen to be mica and hornblende andesites, the ground-masses of which are rarely holocrystalline, but in most cases contain varying amounts of glass. The phenocrysts are generally plagioclases, ranging from andesite to bytownite, orthoclase, microcline, biotite, common green hornblende, and subordinate quartz, as well as accessory apatite and magnetite. The rocks examined were, in most cases, considerably altered to calcite, chlorite, kaolin, and epidote.

These Schwatka andesites occur chiefly as dikes and flows cutting, and overflowing the Tantalus conglomerates and the rocks of the Laberge series. The rocks of the Hutshi group and the Schwatka andesites appear to represent practically contemporaneous intrusions, but no contact between the two was seen, and no facts could be obtained to establish their relative ages. They are both, at least, decidedly older than the Carmack basalts, and newer than the Tantalus conglomerates. It is possible that the Schwatka and Hutshi rocks not only had their origin in the same period of volcanic activity, but that they are also differentiated portions of the same magma, the former being more acidic than the latter. The Schwatka andesites rarely, if ever, contain augite, and generally contain hornblende and biotite phenocrysts; while the feldspars are more acidic than in the andesites of the Hutshi group. The latter always contain augite; but rarely hornblende or even biotite in the first generation, except in very small amounts. So it is not at all impossible that gradations may be found between the two.

The Schwatka andesites are similar to some of the andesites of the Windy Arm series in the Conrad Mining district, and probably belong to the same period of eruption.

Klusha Intrusives.—These intrusives occur prevailingly in the form of stocks or dikes, and are generally seen on the tops of hills and ridges. It is possible, however, that they also exist in the valleys and along the hillsides, covered by the superficial deposits which extend over a large portion of this district. The largest development of them noted is on Porter mountain, on the east side of the Nordenskiöld valley, 14 miles in a direct line from Tantalus. Here, syenite-porphyry covers the greater part of the hill. With this one exception, the exposures of these intrusives are very small, in some cases too small to be accurately shown on maps of the scale of those accompanying this report. The most important of the smaller outcrops—on the summits of North Extension mountain, Contact mountain, and Bunker hill—are shown on the map of the Braeburn-Cynocks coal area.

The Klusha intrusives are syenite-porphyries, generally pink to pale red in colour, and of a coarsely granular habit. They possess, in most places, a typical holocrystalline porphyritic structure, have a fine-grained to aphanatic, microgranitic, quartz-feldspar, ground-mass, and are characterized by the phenocryst combination of alkali-feldspar and lime-alkali feldspar, with subordinate biotite and hornblende.

On Bunker hill are two small exposures of mica syenite-porphyry—a typical porphyritic dike rock, with holocrystalline structure and aphanitic, micrographitic, quartz-feldspar ground-mass, the phenocrysts being andesine, orthoclase, biotite, katoforite, zircon, and magnetite. On Porter mountain is a somewhat similar mica-syenite-porphyry, characterized by large orthoclase phenocrysts, often ½ inch to ¾ inch long, and generally twinned according to the Carlsbad law; the other phenocrysts being chiefly andesine and biotite. Both the plagioclase and the ground-mass are considerably altered. The top of North Extension mountain consists of a finer textured, reddish hornblende-syenite-porphyry, the phenocrysts being chiefly andesine, orthoclase, hornblende, and a little quartz.

On Povoas mountain is a dike of syenite-porphyry, too small to be placed on the map, cutting the rocks of the Hutshi group. In all other places, where outcrops of those Klusha intrusives were seen, they have invaded the Laberge or Tantalus beds.

Carmack Basalts.—These rocks occur chiefly in the northwest portion of the Tantalus coal area. They are developed along both sides of the Lewes river below Tantalus, and extend to the south along both sides of the Nordenskiöld valley for about 10 and 12 miles, on the west and east sides, respectively. They extend to the west past the edge of the Tantalus area; but it was not determined how far. To the east of the Nordenskiöld they outcrop to about the top of the divide between the Nordenskiöld and Lewes rivers; but were nowhere found on the east side of the divide.

These rocks a chiefly grey, bluish, reddish, and brown, to nearly black look basalts and basalt tuffs, and are generally vesicular in structu the cavities being, in places, filled with zeolites.

Along the N. denskiöld river near Carmack's road-house, and at numerous other points, these basalts occur in alternating reddish and greenish layers, are very young looking, and weather and crumble very readily. On Afternoon mountain, about 5 miles southeast of Tantalus, these rocks are almost black in colour; due to the great amount of magnetite they contain. On the top of Shadow mountain, and Deadwood mountain, they alternate with coarser, generally brownish to reddish basalt tuffs, and basalt breccias. Basalt tuffs showing generally distinct bedding, outcrop over nearly the whole west face of Bushy mountain, and are here at least 1,500 feet thick. Bombs up to 18 inches diameter constitute a great part of these tuffs.

When examined microscopically, the ground-mass of these basalts is seen to be generally a mixture in variable quantities of a brown glassy base and augite, plagioclase, and iron ore. The phenocrysts are chiefly bytownite and augite, but small amounts of olivene were also noted in some sections. The chief alteration products are calcite and chlorite.

These Carmack basalts were seen overlying the Tantalus conglomerates, the rocks of the Hutshi group, and the Schwatka andesites, proving them to be newer than these, and consequently later than all the rocks in the district, except the Chisha intrusives. They are probably more recent than these as well, but no contact between the two was seen.

The Carmack basalts are practically the same as those at Miles cañon and to the south of Whitehorse, which have been considered by Dr. G. M. Dawson and others to be of Tertiary age. Hence it is very probable that the Carmack basalts are also Tertiary.

Quaternary.—All the main river valleys in this district are filled with considerable thicknesses of glacial deposits, some of which extend several hundred feet up the hillsides. True boulder-clay spreads over almost the entire length of the upper Lewes valley; high, scarped banks over 100 feet high; of this material overlain by silts, being seen between Lake Laberge and Hootalinkwa, and in other places. Fine silts also occur throughout the length of the Lewes, and on many of its tributaries, including the Nordenskiöld river. They are typically buff coloured with a thickness of 50 to 200 feet, and often rest on a layer of gravel. Along the Lewes river from Whitehorse to Lake Laberge, the white scarped banks, which are often 100 feet high, consist almost entirely of these silts, which are also seen in numerous places along the Lewes below Lake Laberge. On the upper portion of the Lewes, in particular, these silts overlie glacial till, showing them to have been deposited after the retreat of the ice. Overlying these glacial deposits is a thin mantle of recent material which covers the greater part of the whole district.

The Pleistocene and recent terranes of this district are lithologically nearly identical, sometimes grading into each other, and geological studies have, in most cases, not gone far enough to differentiate them. Therefore, on the geological maps to accompany this report, the Quaternary is indicated as a stratigraphic unit.

The Pleistocene deposits near the limits of glaciation are characteristically heavy beds of gravels and sands, which are cross-bedded, and present every indication of having been fluvial deposits in swift water. Away from the source of the material, the deposits become gradually finer, eventually grading into silts, with horizontal bedding, and every indication of sluggish water deposition.

The recent deposits are composed of fluvial, and littoral, sands, gravels, and silts of the present waterways, as well as ground ice, peat, muck, volcanic ash, and soil.

Much of the Pleistocene is directly or indirectly connected with the epoch of glaciation, and though there may be pre-glacial Pleistocene deposits, the definite proof of their occurrence is still wanting

Glaciation has had a very marked effect in shaping the topography in the extreme south of the Yukon Territory, but toward the north as the edges of the former glaciers are reached—it evidently had little effect. In the Braeburn-Kynocks, and Tantalus coal areas, although the glacial ice has widened, and to some extent deepened the valleys and smoothed and rounded the rocky points and ridges, its chief work has been to fill the valleys with great thicknesses of drift material.

On Lake Laberge, both the sides and summits of the rocky hills are heavily glaciated, and to heights of 300 feet above the water the glaciated surfaces of the limestones, where recently stripped of soil and surface covering, are so smooth that to walk over them is difficult. Farther north, glacial striæ were noted to near the top of Shadow mountain, 9 miles south of Tantalus. North of this, markings were not seen, but to the south they are quite plentiful wherever rock is exposed. The directions of the markings at the higher elevations are parallel to the trend of the main north-northwest orographic valleys, but at the lower levels they follow more closely that of the intermediate valleys.

The study of glaciation is of economic as well as scientific interest in this part of the Yukon Territory, as glacial ice is directly connected with the distribution of certain placer gold, and has been the means of covering a number of old streams and river beds that have contained, and do yet contain, rich gold deposits. The greater part of the fine gold found in the bars and banks of the Lewes and its larger tributaries has been washed out of the gravels and superficial deposits through which the present streams have cut their channels since the glacial period. Along the Lewes river for 70 miles below the mouth of the Teslin river, a number of gold-containing gravel-bars have been worked; the Cassiar bar having proved the richest. Limited areas of river flats have also been worked over, where the alluvial cover was not too deep.

A peculiar feature of this district is a layer of volcanic ash or pumiceous sand, which covers the greater part of the country. This was noticed as far south as Lake Bennett, where it is about 1 inch thick. It increases in thickness toward the north and west, and at Lake Laberge is about 5 inches, and at Five Fingers about 11 inches thick. It is very homogeneous, and is more recent than the newest glacial silts. In fact, this ash has fallen since the present waterways have cut out their courses to approximately their present depth, the trees and surface vegetation being rooted in it.

On account of its very even distribution, it appears to have fallen very tranquilly, somewhat like snow, and to have all been deposited continuously; as in it are no intercalated layers of foreign material. It is occasionally, however, found somewhat mixed with other surface deposits where it has been washed from hillsides into the valleys below. Mount Wrangel is the nearest known volcano, and the ashes appear to have come from the direction of that mountain, hence they probably originated either in it, or from some yet undiscovered, extinct volcano in its vicinity. It has been calculated that this ash covers about 25,000 square miles, or has a volume, in all, of at least 1 cubic mile.

A great portion of the entire district considered in this report is covered with superficial deposits obscuring the rock formations over extensive areas, and making geological investigations very difficult. Therefore, on the maps, it has been deemed advisable to distinguish by a separate geological colour, drift covered areas—other than the main valleys which are filled chiefly with Pleistocene deposits—to signify that, on account of this drift, the underlying formations are unknown.

ECONOMIC GEOLOGY.

General Character and Distribution of the Coal.

There are two coal horizons of economic interest in this portion of the Yukon Territory—an upper horizon, occurring near the top of the Tantalus conglomerate, to which belong the seams at the Tantalus mine and at Tantalus butte; and a lower horizon some distance below the coal conglomerates, which includes the seams at the Five Fingers mine, to the west of the 69 mile-post from White-horse, on the Whitehorse-Dawson road, and elsewhere. In consider-ing the distribution of the coal it might be said, in a general way, that it is chiefly to be found in the areas covered by the Tantalus conglomerates. It would, however, be quite possible for coal seams of the lower horizon to be found where the overlying conglomerate was not to be seen, being either covered by other formations or deposits, or having been eroded away. Still, in only one place in the district, namely, at the Five Fingers mine, have coal seams of any economic value been found where the Tantalus conglomerates are not in evidence. In any case, the Coal Measures of the upper horizon contain much the more valuable coal seams. Also, in all geological, or prospecting work in this district, those conglomerates form a very valuable geological horizon marker, which is very readily identified, and, when found, the approximate positions of both coal horizons can be determined at once.

The Coal Measures of the upper horizon, in all probability ex-tend to the north along the Tantalus Butte ridge, to about 2½ miles from the Lowes river, and perhaps much farther; although to the north of this the sedimentary formations are covered with basalts and superficial deposits. In the valley to the east of Tan-talus butte the lower horizon is several hundred feet below the sur-face, but probably outcrops along the river banks farther up the river. However, no coal seams more than 2 to 3 inches thick were seen.

To the south of Tantalus, there is believed to be an almost con-tinuous belt of the Tantalus conglomerates along the east side of the Nordenskiöld valley—for about 17 miles; but they are covered in most places by volcanic rocks and superficial deposits. Underlying, or contained in the conglomerates, the coal seams of the lower and upper horizons, respectively, may be expected to occur, the latter being possibly, however, in some places eroded away. The seams belonging to it were nevertheless seen in several places.

About one-fourth mile above the 114th mile-post (Tantalus being at 131) and one-fourth mile east of the wagon road on the south side of Porter mountain, a portion of the upper measures is exposed, the upper part of which has been eroded. The measures here are considerably folded and distorted by the intrusion of the syenite-porphyry, which outcrops over most of Porter mountain, and the quality of the coal is changed from that of a lignite, or bituminous coal, as elsewhere, to an anthracite. Some prospecting work has been done on this coal, which is locally known as ' Porter's coal.' The widest seam found was only about 16 inches, this being the one on which the work had been done. However, other and better seams may exist, as the coal-bearing beds are mostly heavily drift covered. An average outcrop sample of this 16 inch seam, analysed by Mr. F. G. Wait, of the Mines Branch, Department of Mines, was found to contain :—

Water.	Vol. Combustible Matter.	Fixed Carbon.	Ash.
4·68	15·59	72·26	7·47

Farther south, coal of the upper horizon was also seen along the wagon-road about opposite the 118 mile-post, and west of Lone Pine mountain; but the thicknesses of the seams were not ascertained, as the superficial covering is very heavy.

The Tantalus conglomerates outcrop on a small round hill facing the Lewes river, about 1 mile from Tantalus, dipping to the west as on Tantalus butte. Here the coal seams of the upper horizon exist, in all probability, near the top of the hill. However, as no prospecting has been done, and the superficial deposits are in most places heavy, no coal—up to the present time—has been found.

In the southwest portion of the Braeburn-Cynocks coal area, the Tantalus conglomerates are quite extensive, and overlie conformably the beds of the Laberge series in the form of a large, broken, flat, undulating cake, which forms the top, and, in places, the main portions of a number of hills and ridges. (For a description of the area covered by the Tantalus conglomerates, see the chapter on General Geology). Here the upper portion of the Tantalus conglomerates—including the upper coal horizon—has been, in most places at least, eroded away. But since the lower horizon exists in the upper Laberge beds, generally only about 200 or 300 feet below the Tantalus conglomerates, it follows that this horizon underlies all the conglomerate beds in the locality, and would outcrop in the valleys or along the hillsides around the edges of the conglomerate areas, wherever the upper beds of the Laberge series are not below the valley bottoms, were it not that it is in most cases covered by volcanics and superficial deposits.

The coal measures of this lower horizon were seen outcropping along the northeast face of Red ridge, and are well exposed along a small creek on the northeast side of Division mountain. Here one seam is 7 to 8 feet; one is 4 feet, and several are from 6 to 18 inches wide. Nos. 1 and 2 are average surface samples taken from an 18 inch, and the 7 to 8 ft. seams, respectively. They were assayed by Mr. F. G. Wait of the Mines Branch, Department of Mines, and gave the following:—

Water.	Volatile Combustible Matter.	Fixed Carbon.	Ash.	Fuel Ratio.
No. 1— 8·98	29·62	48·30	13·10	1 : 1·63
No. 2—12·02	34·28	42·56	11·14	1 : 1·24

These coals are classed by Mr. Wait as lignites. Other coal seams may exist here which are covered, and were not seen, and these measures may also yet be found outcropping in other portions of this Braeburn-Cynocks area in the vicinity of the Tantalus conglomerate, where the superficial deposits have been removed, as is often the case on steep sidehills or in creek-bottoms.

The hills along the left side of the Lewes river, commencing 10 miles below Hootalinkwa and extending down stream 5 miles, to below the mouth of Fife creek, also consist chiefly of the Tantalus conglomerates. Here, the lower coal horizon must underlie all this conglomerate area, and careful prospecting may even result in finding one or both of the coal horizons outcropping at the surface. This part of the country is in most places heavily covered with both fallen, and standing timber, and the consolidated rock terrances are, to a great extent, hidden by considerable thicknesses of superficial deposits, so that prospecting is rendered very difficult.

DESCRIPTION OF MINING PROPERTIES.
Tantalus Mine.

The Tantalus mine is situated on the left limit of the Lewes river about 190 miles down the river from Whitehorse. It is about 2 miles, in an easterly direction, from Carmack's road-house, on the Whitehorse-Dawson wagon-road, 131 miles from Whitehorse. A police-barracks, a store, and the Carmack post-office, are situated midway between the mine and the road-house.

The coal outcrops in the river banks and is, therefore, well situated for economical working. The cars are hauled out of the adits by mules, and by means of a cable, operated by a small stationary steam engine, are pulled up an incline, at the top of which the coal is dumped into bunkers ready for loading into river boats. Three seams have been opened up, only the lower two of which have been worked to any extent. The seams are somewhat variable in width, but have averaged, perhaps 7'-6", 6'-6", and 3 feet of coal in the bottom, middle, and top seams respectively. The lower two seams have, in places, not more than 4 feet of rock between them, and the middle and top seams are generally about 7 feet apart. The coal is worked by the stall and pillar system, from two adits, which, in September, 1907, had been driven about 1,800 feet. From the bottom seam, nine rooms had been, or were being worked, eight of which were from 50 to 115 feet up, while No. 1 had been run 160 feet to the surface for air. From the middle seam, 23 rooms had been opened up. The dips are to the east and vary in the workings from 24° to 40°. The seams are somewhat dirty, but the coal could easily be washed or stored. This is not done, possibly because wages here are high. Miners get $1.50 per car, holding each about 2,500 pounds, for mining the coal; and after paying for board, powder, oil, caps, etc., they make from $5 to $6 per day.

The three seams are carefully sampled. A and B being average samples of the bottom and middle seams, respectively, taken from the breasts of the adits, when in about 700 feet; while C is an average sample of the top seam where it is cross-cut from the middle

seam, 300 feet from the adit entrance. These were analysed by Dr.
Hoffmann, formerly of the Geological Survey, and gave the following
results:—

Sample.	A.	B.	C.
Water,	0 75	0 76	0 82
Volatile combustible matter	23 61	24 74	25 12
Fixed carbon	55 21	58 60	66 03
Ash	20 43	15 90	8 03
	100 00	100 00	100 00
Firm coherent coke per cent	75 64	74 50	74 06

As these coals—in the laboratory—showed a good quality of
coke, it is hoped that they will, in practice, produce a coke suitable
for smelting purposes.

There is, at present, only a small demand for this coal in Dawson and Whitehorse; it is mainly used for fuelling some of the river
steamers. Tantalus, and possibly Tantalus butte and the Five
Fingers mine, are the only points along the river between Whitehorse and Dawson—a distance of 460 miles—at which coal is
obtainable, consequently to use coal on the boats exclusively, would
necessitate carrying a sufficient supply from Tantalus to last the
trip to either Whitehorse or Dawson, and return. On this account,
wood, obtainable in many places along the river, is still the chief
fuel used. But, the supply of wood accessible to the river is not
great, is fast diminishing, and will only last at most a few years
more, when coal or oil will have to be used. Oil from the United
States is used for fuelling the steamers on the Yukon river below
Dawson.

The output of the Tantalus mine in 1906 was 5,173½ tons, and
in 1907 was nearly 10,000 tons. In 1908 a much smaller amount
was mined.

Tantalus Butte.

On Tantalus butte, across the river from the Tantalus mine, the
same Coal Measures again outcrop; but here dip to the west. The
coal outcrops are near the top of the butte, and generally 300 or 400
feet above the river, superficial deposits covering the formation
lower down. A partial section of the measures exposed here gave as
follows:—

Heavy massive conglomerate beds, at least 200 feet thick, forming top of the butte.

	Feet.
Shales, sandstones, etc., including an 8'-10" seam of coal...	70
Conglomerate..	60
Shales, sandstones, etc., including a 9'-10" seam of coal near the top, and a 7 ft. seam near the bottom: total thickness..	160
Conglomerate..	80
Finer sandstones, shales, etc...........................	100
Conglomerate, bottom not seen.	

Average outcrop samples of the 8'-10"; the 9'-10"; and the best 6 feet of the 7 ft. seam, numbered respectively, A, B, and C, were assayed by Mr. F. G. Wait of the Mines Branch, Department of Mines, and gave the following results:—

	A.	B.	C.
Water......................................	13·64	16·32	12·87
Volatile combustible matter.............	31·83	31·72	31·72
Fixed carbon..............................	51·84	42·13	49·51
Ash.......................................	2·69	9·83	5·90
	100·00	100·00	100·00
Ratio of volatile combustible matter to fixed carbon.................................	1·63	1·33	1·56
Potash reaction..........................	Dark.	Brownish.	Red.
Colour of ash............................	Pale reddish brown.	Pale brownish yellow.	Yellowish brown.
Kind of fuel..............................	Lignite.	Lignite.	Lignite.

These do not give a coherent coke.

When these Tantalus Butte seams have been developed, and samples can be obtained at a sufficient distance from the surface to be free from all weathering agencies, it is expected that the coal will be very similar to that at the Tantalus mine. With the exception of a small amount of trenching and stripping, however, no work as yet has been done on this property.

The Five Fingers Mine.

This property is situated on the right bank of the Lewes river, 16 miles below Tantalus, by the river, or 8 miles distant, in a direct

line, in a direction a few degrees west of north. Some years ago a
slope was sunk about 350 feet—and rooms driven off it—on the best
seam so far found in these measures, and which dips at 16° to the
east; the seam in the lower rooms being 3½ to 4 feet thick. A considerable amount of coal was mined and sold, chiefly in Dawson, but
the workings have now been closed for several years.

As the top of this old slope is subject to mud slides—being
situated in the steep clay and sand bank of the river—when work
was recommenced in 1906, under new management, it was on safer
ground—some distance to the south. Here a new slope was sunk
783 feet, on a seam higher in the measures than the seam in the old
workings, and which also dips at 16° to the east. This seam—which
in places in the slope is not more than 6 inches thick—at the bottom contains 22 inches of good clean coal, and 24 inches of coal and
shale.

During 1907 and 1908 very little work was done on this property.
In the former year a 26 ft. winze was sunk at 450 feet down the
new slope, to a seam of coal 4′-6″ thick, which is apparently the same
seam as that in the old slope.

Sample A is an average of the 22 inches of good coal in the bottom of the 783 feet slope; and B is an average sample of the bottom
of the 26 ft. winze. These were assayed by Mr. F. G. Wait of the
Mines Branch, Department of Mines, and gave the following
results:—

Sample.	A.	B
Water	5·95	5·29
Volatile combustible matter	40·46	36·14
Fixed carbon	45·16	40·12
Ash	8·43	18·45
	100·00	100·00
Coke per cent	53·59	58·57
Character of coke	Firm coher-ent.	. . . .
Ratio of volatile combustible to fixed carbon	1 to 1·11	1 to 1·11
Colour of ash	Reddish.	Reddish.
Kind of fuel	Coal.	Coal.

Plate VIII.

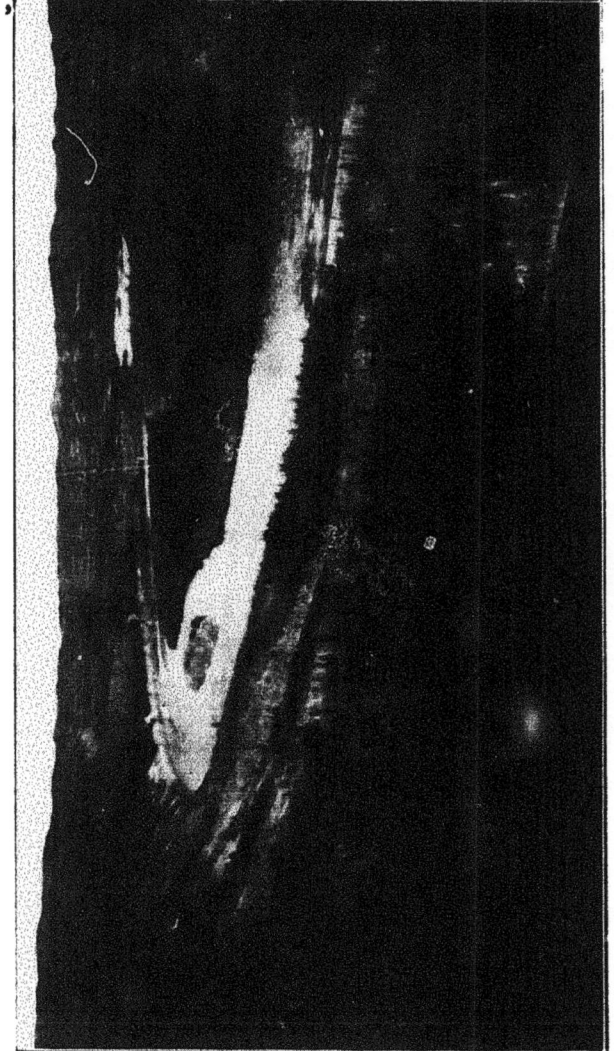

The Lewes river above Tantalus.

APPENDIX I.

Mack's Copper.

During the season of 1908, in addition to the work in the Tantalus and Braeburn-Cynocks coal areas, the writer made an excursion up the Hutshi river, and visited some groups of mining claims known locally as 'Mack's Copper' and the 'Gilltana Lake Claims.' The former is situated only a few miles to the southwest of Montague, and a short distance west of the western edge of the Tantalus map, and is so named because originally owned by the Mack brothers.

This property is usually reached by following the Whitehorse-Dawson road to about 6 miles above Montague, where a branch road turns off to the southwest, following approximately the old Dalton trail up the Hutshi river. This branch road may be followed about 8 miles to where a trail ascends the hills northward to the claims, which are practically on the summits, at about 4 miles from, and 1,000 feet above, the valley of the Hutshi river.

Practically all the ore to be seen in the vicinity appears to be on one claim, and occurs in a fine-grained, greenish, andosite, at or near its contact with limestone. The ore consists chiefly of magnetite, with hematite in minor quantities, both more or less impregnated with copper minerals, chiefly chalcopyrite, malachite, and azurite. The main mass of mineral is in the form of a small hill of almost solid iron ore, about 200 feet wide, and from 300 to 400 feet long.

On the top of the hill generally there are no indications of copper, this having, possibly, been leached out and redeposited lower down; but over the south side of the hill the iron ore carries considerable copper. A prospecting drift had been started in one of the most promising looking places in the hillside, and when visited in July, 1908, had been driven 38 feet. The only other work done on this property was in the form of an open-cut on the adjoining hill to the west—the ore in the two places apparently not being connected. At the last-mentioned place the ore is only 10 feet to 12 feet wide, and is lying next to, and more or less in, the limestone. The richest ore is found here, but it has only been traced a short distance. It may continue, however, farther along the limestone, the contact being so deeply covered with superficial deposits, that only a few feet have been uncovered.

The following samples were taken by the writer, and have been assayed by Robt. Smart, Government Assayer at Whitehorse.

No. 1 is an average of the end of the drift.

No. 2 is an average of the best 4 feet in the open-cut.

Sample.	Gold.		Silver.		Copper.	Total Value.
	Ozs. per ton.	Val. per ton.	Ozs. per ton.	Val. per ton.	Percentage.	Per ton.
1	Trace.	...	Trace.	...	1·80	$ 4·68
2	0·025	$1·51	3·4	$1·87	5·55	16·81

APPENDIX II.

The Gilltana Lake Claims.

After visiting ' Mack's Copper ' (See Appendix I) the writer
continued along the Dalton trail up the Hutshi river to Hutshi
lake and village situated about 50 miles from the Whitehorse-Dawson wagon-road; thence to Gilltana lake, a distance of 15 miles in a
northwesterly direction. t'lnims have been staked here on both sides
of the lake, and are generally known as the ' Gilltana Lake claims.'
The greater part of the staking was done during the summer of 1907,
but a number of new discoveries were made during the following
season.

The ore on the northwest side of the lake occurs at the contact
between granite and limestone, and is in the form of narrow lenses
of mineralized matter and quartz, the widest lens seen being about
4 feet wide. Generally, these are only from 1 to 2 feet wide, and
have no present economic value.

On the northeast side of the lake the claims are chiefly located
over the face of a hill which rises to about 1,200 feet above the lake.
The formation here is chiefly mica schist, interbanded with which
are beds of quartzite and limestone, the latter being, generally, only
3 to 4 feet thick, but in places, as thick as 50 feet. The formation
strikes about parallel to the lake, and the beds dip into the hill at
low angles; so that the different schist bands extend along the face
of the hill one above the other, maintaining an almost horizontal
outcrop. In places these bands are mineralized with magnetite,
generally carrying copper minerals—chiefly chalcopyrite and malachite—these mineralized bands constituting the ore-bodies. The
original schists show all degrees of mineralization and replacement,
from being non-metalliferous, to consisting of almost solid iron ore.
The best of these mineralized bands or zones generally average from
6 to 10 feet in width; although one of 20 feet was seen exhibiting
intense mineralization. These can be traced generally from 50 to
100 and even 200 feet; when the iron and copper minerals become
gradually less and disappear, or, at times, follow other parallel bands.

Three prominent, and other less important bands were observed at different elevations along the face of the hill.

On the Helen claim, up Franklin creek, where some open-cut work has been done, streaks of copper ore much richer than observed elsewhere, and 1 to 3 feet thick, were, in places, seen included in wider bands of less highly mineralized matter.

The formations have been cut by dikes of light coloured hornblende and mica andesites, and dark fine-grained basalt, but these have had no visible effect on the ore deposits which are apparently genetically connected with the intrusive granites in the vicinity. The district is well worth prospecting and a number of the claims look very promising indeed.

Sample No. 3 is an average taken at the surface across one of the best looking 6 ft. bands.

Sample No. 4 is an average across one of the best streaks 3 feet wide on the Helen claim.

No.	Gold.		Silver.		Copper.	Total Value.
	Ozs. per ton.	Val. per ton.	Ozs. per ton.	Val. per ton.	Percentage.	Per ton.
3	Trace.	. . . . . .	Trace.	. . . . . .	1 35	$ 3 51
4	Trace.	. . . . .	Trace.	. . . . . .	9·00	23.40

APPENDIX III.

Coking Tests on Coal from Tantalus Mine.

(Samples of coal were submitted to Dr. Eugene Haanel. Tests made by E. Stansfield, M.Sc., in the Otto Hoffmann Coke Ovens of the Dominion Iron & Steel Co., Ltd., at Sydney, N.S.)

The method employed was as follows: a suitable amount of the coal sample was weighed out, damped until it contained about 5 to 10 per cent of water, and charged into a sheet iron box having a perforated lid. The lid was left loose so that it could sink into the box, thus ensuring the coal tested receiving the full pressure of the weight of the coal lying above it in the oven. This box was then put into an empty coke oven, which was ready for charging, in such a way that it rested on the floor of the oven, some little way from the end, and occupied almost the full width of the oven. All the six boxes of coal tested were placed together in the same oven, which was then charged with moist, washed coal exactly as in the ordinary practice of the Steel Company. The oven was coked as usual, but was not pushed until forty-eight hours had elapsed after charging. As soon as the coke was pushed the test boxes were recovered from amongst the hot coke and carefully quenched. The boxes were dried for a day over a hot flue before weighing and opening.

A series of preliminary experiments showed that the box test produces a coke which is closely comparable with the coke produced from the same coal on the large scale.

Description of Cokes Produced.

Coal from Upper Seam— ... ... d throughout, but the product was a dirty gr.. ... ... crumbly, like dried mortar in appearance. No reg ... ...

Not a commercial coke.

Yield from dry coal was 75·9 per cent.

Coal from Upper Seam—Washed.—Coke was much better and sounder than that from the raw coal. Less crumbly, though somewhat similar in appearance. Very dense, breaks cleanly.

A poor commercial coke.

Yield from dry coal was 75·3 per cent.

Coal from Middle Seam—Raw.—Product similar to that from the Upper Seam raw coal. More like hard mortar than coke. Very dense and contains a lot of dirty spots. No regular cleavage.

Not a commercial coke.

Yield from dry coal was 75·8 per cent.

Coal from Middle Seam—Washed.—Coke not so good as from the washed coal of the Upper Seam, but a great deal better than from the raw coal of the Middle Seam. Very dense, not much breeze produced in breaking, but fracture not very regular.

A possible commercial coke.

Yield from dry coal was 77·4 per cent.

Lower Seam—Raw Coal.—A sound coke produced throughout the box. Clean fracture, but breaks rather easily in any direction. Could probably be used in a blast furnace for smelting iron, although there is very little cellular structure.

A very fair commercial coke.

Yield from dry coal was 74·6 per cent.

Lower Seam—Washed Coal—Coke like that from the corresponding raw coal, not noticeably cleaner, but harder and sounder.

A commercial coke.

Yield from dry coal was 74·1 per cent.

The six samples were all very fine powder, containing only a few small lumps. If the coal had been rather larger, it is probable that a more cellular coke would have been obtained.

The coals hardly shrank at all on coking.

Washing Tests on Coal from the Tantalus Mine.

Tests made by H. G. Carmichael, B. Sc., in the Mining Laboratories of McGill University, Montreal, Que.

The three samples of coal were submitted to heavy solution separations, and were also washed on a small scale in an experimental apparatus.

The following is the summary of the results obtained.

COAL: UPPER SEAM.

—	Yield.	Ash Contents.
	,	,
		7.
HEAVY SOLUTION SEPARATIONS —		
Good coal (i.e. under 1·375 specific gravity)....	38·0	4·5
Bone coal (between 1·375 and 1·55 sp. gr.)........	40·0	14·2
Refuse (over 1·55 sp. gr.)...	22·0	44·1
Good and bone coal together (under 1·55 sp. gr.)..	78·0	9·5
Original coal...........	... .	17·0
WASHING TRIALS—		
Ash in raw coal.....................	17·0	
Ash in washed coal	13·8	
Yield of washed coal.	81·0	
Reduction in ash	18·8	
Calorific value of raw coal....	6790 cals. per gram	
" " washed coal	7110 " "	
Increase in calorific value..	6·1"	

COAL: MIDDLE SEAM.

—	Yield.	Ash Contents
		. ,
		. ,
HEAVY SOLUTION SEPARATION—		
Good coal (i.e. under 1·375 sp. gr.).............	23·0	5·2
Bone coal (between 1·375 and 1·55 sp. gr.).......	50·0	11·7
Refuse (over 1·55 sp. g.)...	27·0	25·1
Good and bone coal together (under 1·55 sp. gr.)..	73·0	11·7
Original coal (calculated from above figures)......	... .	21·0
WASHING TRIALS—		
Ash in raw coal	19·2	
Ash in washed coal.	11·0	
Reduction in ash...	27·0	
Calorific value of raw coal	6310 cals. per gram.	
" " washed coal............	7070 " "	
Increase in calorific value...	12·1"	
Yield of washed coal....	76·5"	

COAL: LOWER SEAM.

—	Yield.	Ash Contents.
	%	%
HEAVY SOLUTION SEPARATOR—		
Good coal (i.e. under 1·375 sp. gr.)	53·0	5·3
Bone coal (between 1·375 and 1·55 sp. gr.).... .	25·0	15·3
Refuse (over 1·55 sp. gr.)...................	22·0	40·0
Good and bone coal together (under 1·55 sp. gr.)..	78·0	8·5
Original coal	....	16·2
WASHING TRIALS—		
Ash in raw coal...	16·2	
Ash in washed coal	12·7	
Yield of washed coal....	83·0	
Reduction in ash ... ...,..............	21·6	
Calorific value of raw coal..	6790 cals. per gram.	
" washed coal...........	7210 "	
Increase in calorific value..	6·2%	

It can be seen from the figures given for the heavy solution separations, that a considerable reduction in ash can only be obtained in conjunction with a small yield of washed coal; a large part of the ash forming impurities being intimately bound up with combustible matter.

The coals from the Upper and Lower seams are especially unsuited for washing. The coal from the Middle seam is so dirty that it might be profitable to wash it.

Chemical Tests on Coal from the Tantalus Mine, White Pass and Yukon Railway Company, made by the Mines Branch.

	UPPER SEAM.		MIDDLE SEAM.		LOWER SEAM.	
	Raw	Washed	Raw	Washed	Raw	Washed
	%	%	%			%
Moisture in sample as received in laboratory..........	0·9	....	0·7	....	0·7	....
Proximate Analysis of Coal Dried at 105°C —						
Fixed carbon.............	58·0	59·9	54·1	60·3	56·0	59·2
Volatile matter......	25·0	26·3	26·7	25·7	27·8	28·1
Ash...............	17·0	13·8	19·2	14·0	16·2	12·7
Ultimate Analysis of Dried Coal -						
Carbon........	69·8	.....	..		71·1	.....
Hydrogen......	4·0	....	....	....	4·3	....
Sulphur..	0·5	0·5	0·5	0·4	0·5	0·5
Nitrogen	0·8	0·8	0·9	0·8	0·7	0·8
O₂	7·9		... ...	....	7·2	....
Ash	17·0	... .	....	16·2	.....

| Calorific value of dried coal in calories per gram ... .. | 6,700 | 7,110 | 6,310 | 7,070 | 6,750 | 7,210 |

INDEX

10190—5

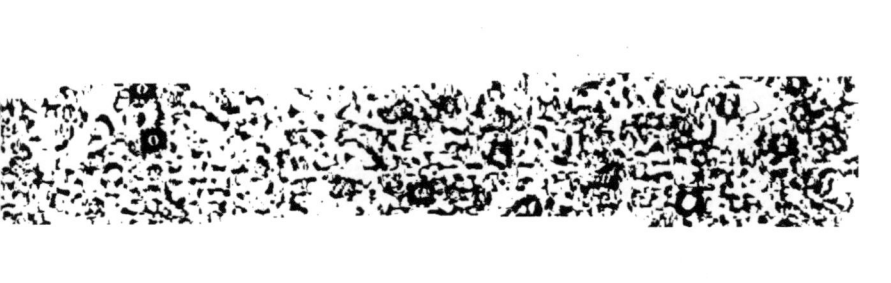

MICROCOPY RESOLUTION TEST CHART

(ANSI and ISO TEST CHART No. 2)

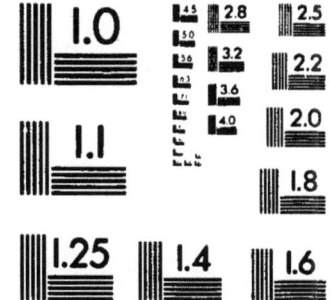

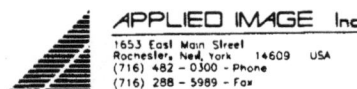

APPLIED IMAGE Inc

1653 East Main Street
Rochester, New York   14609   USA
(716) 482 - 0300 - Phone
(716) 288 - 5989 - Fax

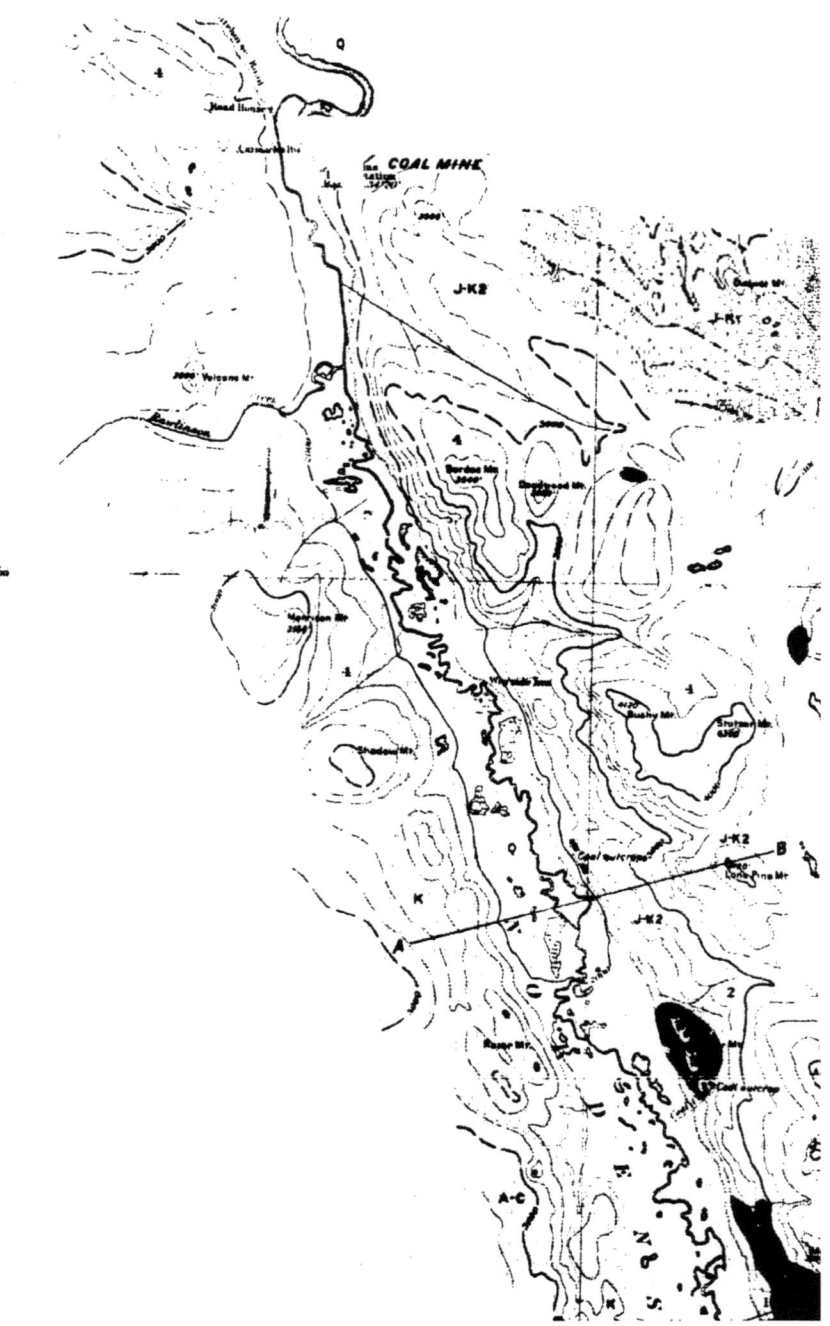

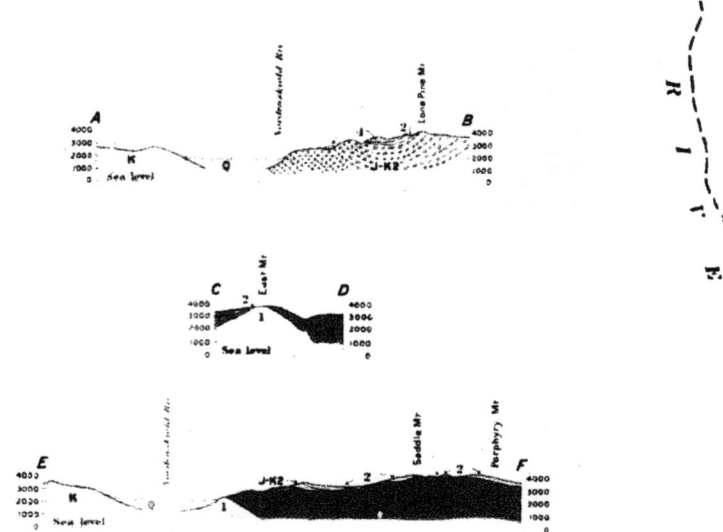

Geological sections along the lines *AB. CD. EF.*

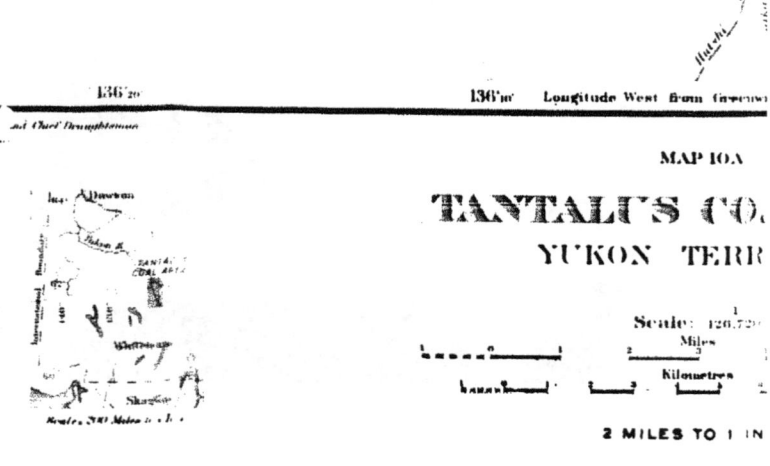

TANTALUS CO.

YUKON TERR

Scale: 126,720

2 MILES TO 1 IN

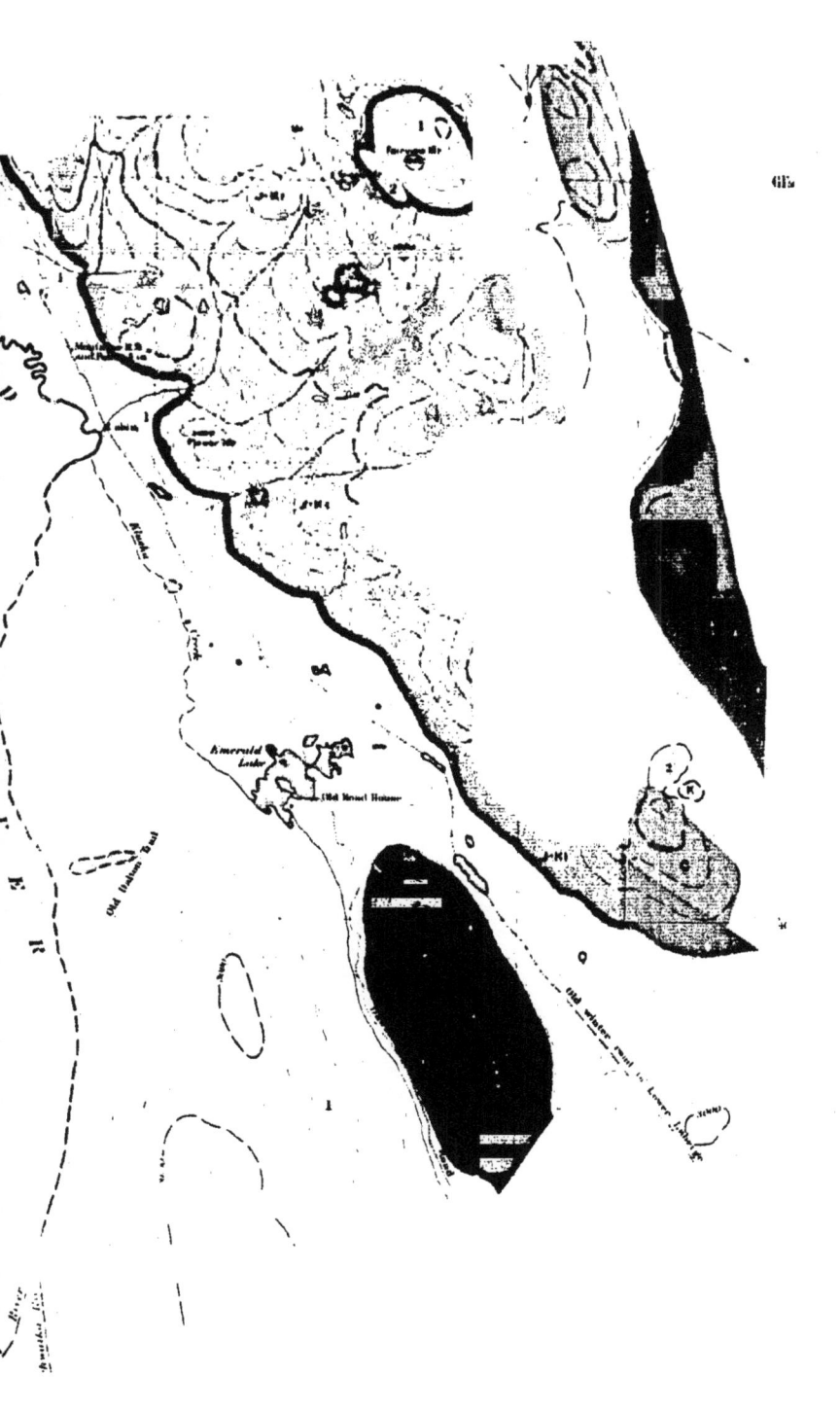

Lightning Source UK Ltd.
Milton Keynes UK
UKHW021007161218
334046UK00008B/747/P